**Recent Development of Aerodynamic Design Methodologies**

Edited by
Kozo Fujii and
George S. Dulikravich

# Notes on Numerical Fluid Mechanics (NNFM)    Volume 68

Series Editors: Ernst Heinrich Hirschel, München (General Editor)
Kozo Fujii, Tokyo
Werner Haase, München
Bram van Leer, Ann Arbor
Michael A. Leschziner, Manchester
Maurizio Pandolfi, Torino
Arthur Rizzi, Stockholm
Bernard Roux, Marseille

Volumes 1 to 51 are out of print.
The addresses of the Editors are listed at the end of the book.

# Recent Development of Aerodynamic Design Methodologies

– Inverse Design and Optimization –

Edited by
Kozo Fujii and
George S. Dulikravich

vieweg

http://www.vieweg.de

Produced by Geronimo GmbH, Rosenheim
Printed on acid-free paper

ISSN 0179-9614

ISBN 978-3-322-89954-5          ISBN 978-3-322-89952-1 (eBook)
DOI 10.1007/978-3-322-89952-1

Susumu Takanashi

## Dedication

Dr. Susumu Takanashi, the former Computational Fluid Dynamics (CFD) researcher at the National Aerospace Laboratory (NAL) in Japan, passed away on October 28, 1995, at the age of 56. He was engaged in developing CFD design methods at the NAL for many years. The aerodynamic shape inverse design method that he developed has been and still is used by many researchers and practicing engineers in the world.

Dr. Takanashi's results were used in the aerodynamic design process of the YXX aircraft and others in the 1980's. Recently, the design method was used for the modification of the National Space Development Agency (NASDA) HOPE-X space vehicle configuration that may fly early in the year 2000. In addition, his inverse design method is a main aerodynamic tool for the experimental supersonic transport (SST) aircraft, planned to fly early in 20002 by the NAL.

Dr. Takanashi contributed significantly to the development of the CFD analysis software at the NAL. His contribution was not restricted to the development of CFD design methods. He led the group of CFD analysis code developers for three-dimensional Navier-Stokes flow-field simulations over wings, wing-fuselage combinations, and full aircraft configurations. At the same time, he developed a three-dimensional block-structured computational grid generation code based on his own idea. The code became an important part of the simulation of the transonic flow fields around aircraft configurations. He also contributed to the introduction of the Numerical Wind Tunnel (NWT) at the NAL, still the largest practical supercomputer system in the world with more than 220 GFLOPS peak performance. The NWT is now playing the key role in the aerodynamic analysis and design of the HOPE-X and NAL's SST supersonic experimental research vehicle.

Dr. Takanashi worked together with many CFD researchers as shown by his technical publication list. He influenced his co-workers profoundly, not only from the point of view of how to do the advanced research, but also with his attitude toward the research. We all miss him and will remember him as our teacher, fellow researcher, and a friend that could have offered many more new exciting contributions to the profession.

As a humble token of our deep appreciation and respect for the memory of Dr. Susumu Takanashi, we prepared this volume on Aerodynamic Shape Inverse Design and Optimization, and would like to dedicate it to him. Listed below are his resume and a list of technical papers memorizing his contribution.

K. F.
G. S. D.

**Susumu Takanashi**

1939    Born on March 6th, in Chiba, Japan
1958    Employee of Muromachi-Kaiun
1960    Technical Assistant, National Aerospace Laboratory
1962    Research Scientist, National Aerospace Laboratory
1963    Graduated from University of Electro-Communications, Junior
        College, March 1963
1968    Graduated from Science University of Tokyo, Department of Physics,
        March 1968
1975    Received ph. D. from University of Tokyo, December 1975
1977    Principal Research Scientist1985 Section Chief, Aircraft
        Performance Research Section
1988    Section Chief, Computational Aerodynamics Research Section
1995    Died October 28th at the age of 56

## Representative  Research  Papers

Takanashi, S., "A Method of Obtaining Transonic Shock-Free Flow around Lifting Aerofoils", Transactions of the Japan Society for Aeronautical and Space Sciences, Vol. 16, No. 34, pp. 246-263, Dec. 1973.

Takanashi, S., "An Iterative Procedure for Three-Dimensional Transonic Wing Design by the Integral Equation Method ", AIAA Paper 84-2155, Aug., 1984.

Tatsumi, S. and Takanashi, S., "Experimental Verification of Three-Dimensional Transonic Inverse Method", AIAA Paper 85-4077, Oct., 1985.

Hirose, N., Takanashi, S. and Kawai, N., "Transonic Airfoil Design Based on Navier-Stokers Equations to Attain Arbitrarily Specified Pressure Distribution  − An Iterative Procedure", NAL TR-901T, March 1986.

Obayashi, S., Fujii, K. and Takanashi, S., "Toward the Navier-Stokers Analysis of Transport Aircraft Configurations", AIAA Paper 87-0428, Jan. 1987.

Sawada, K and Takanashi, S., "A Numerical Investigation on Wing/Nacelle Interferences of USB Configuration", AIAA Paper 87-0455, Jan. 1987.

Miyakawa, J., Takanashi, S., Fujii, K. and Amano, K., "Searching the Horizon of Navier-Stokers Simulation of Transonic Aircraft", AIAA Paper 87-0524, Jan. 1987.

Hirose, N., Takanashi, S. and Kawai, N., "Transonic Airfoil Design Procedure Utilizing a Navier-Stokers Analysis Code", AIAA Journal, Vol. 25, No. 3, pp. 353-359, March 1987.

Takanashi, S., Obayashi, S., Matsushima, K. and Fujii, K.,

"Numerical Simulation of Compressible Viscous Flows around Practical Aircraft Configurations", AIAA Paper 87-2410, Aug. 1987.

Hirose, N and Takanashi, S., "Some Topics in Computational Transonic Aerodynamics", NAL TR-1018T, April, 1989.

Matsushima, K., Takanashi, S. and Fujii K., "Navier-Stokers Computations of the Flows about a Space-Plane", AIAA Paper 89-3402, Aug. 1989.

Takanashi, S., "Large-Scale Numerical Aerodynamic Simulations for Complete Aircraft Configurations", NAL TR-1073T, July 1990.

Takanashi, S., "A Simple Algorithm for Structured-Grid Generation with Application to Efficient Navier-Stokers Computation", Computers & Fluids, Vol. 19, No. 3/4, pp. 393-399, 1991.

Sudani, N., Kaneda, H., Sato, M., Miwa, H., Matsuo, K. and Takanashi, S., "Evaluation of NACA0012 Airfoil Test Results in the NAL Two-Dimensional Transonic Wind Tunnel", NAL TR-1109T, May. 1991.

Fujii, K. and Takanashi, S., "Aerodynamic Aircraft Design Methods and Their Notable Applications - Survey of the Activity in Japan", Third International Conference on Inverse Design Concepts and Optimization in Engineering Sciences (ICIDES-III), ed.: G.S. Dulikravich, Washington, D.C., Oct. 23-25, 1991.

Kaiden, T., Ogino, J. and Takanashi, S., "Non-Planar Wing Design by Navier-Stokes Inverse Computation", AIAA Paper 92-0285, Jan. 1992.

Takanashi, S. and Takemoto, M., "A Method of Generating Structured-Grids for Complex Geometries and Its Application to the Navier-Stokers Simulation", Computational Fluid Dynamics Journal, Vol. 2, No. 2, pp. 209-218, July 1993.

Takanashi, S., "Numerical Simulation of High Incidence Flow over a Space-Plane at Supersonic Speed", Fluid Dynamics of High Angle of Attack , R. Kawamura and Y. Aihara (Eds.), pp. 339-350, Springer-Verlag, 1993.

Matsushima, K. and Takanashi, S., "Navier-Stokers Simulation of Transonic Flows about a Space-Plane", AIAA Paper 94-1864, June 1994.

Takanashi, S. and Takemoto, K., "An Automatic Grid Generation Procedure for Complex Aircraft Configurations", Computers & Fluids, Vol. 24, No. 4, pp. 393-400, 1995.

# Preface

Computational Fluid Dynamics (CFD) has made remarkable progress in the last two decades and is becoming an important, if not inevitable, analytical tool for both fundamental and practical fluid dynamics research. The analysis of flow fields is important in the sense that it improves the researcher's understanding of the flow features. CFD analysis also indirectly helps the design of new aircraft and/or spacecraft. However, design methodologies are the real need for the development of aircraft or spacecraft. They directly contribute to the design process and can significantly shorten the design cycle. Although quite a few publications have been written on this subject, most of the methods proposed were not used in practice in the past due to an immature research level and restrictions due to the inadequate computing capabilities.

With the progress of high-speed computers, the time has come for such methods to be used practically. There is strong evidence of a growing interest in the development and use of aerodynamic inverse design and optimization techniques. This is true, not only for aerospace industries, but also for any industries requiring fluid dynamic design. This clearly shows the matured engineering need for optimum aerodynamic shape design methodologies. Therefore, it seems timely to publish a book in which eminent researchers in this area can elaborate on their research efforts and discuss it in conjunction with other efforts.

With this as a background, we have decided to prepare this book entitled "Recent Development of Aerodynamic Design Methodologies – Inverse Design and Optimization -". All the contributing authors are well-recognized researchers in this field. A different author covering another aerodynamic shape design methodology writes each chapter.

Three categories of design methodology are considered: Genetic Algorithms, Inverse Design, and Optimization. "Genetic Algorithms (GA)" are rapidly gaining popularity and may become the methods of choice for multi-objective and interdisciplinary optimization. "Optimization" can be considered as generalized design, which introduces integral target properties and constraints. "Inverse Design" describes methods to find a configuration that realizes, for instance, the target pressure distribution. Although this approach has some drawbacks, such as the difficulty of finding good target pressure distribution, it has been accepted and is practically used by industry.

In addition, two short contributions are added from Japanese industries. These contributions describe how they used Dr. Takanashi's inverse design method in their practical applications.

As editors of this book, we would like to acknowledge all the contributing authors for their effort and patience in the process of preparing this publication. We also would like to thank Prof. Ernst H. Hirschel, the general editor of the series of Notes on Numerical Fluid Mechanics, as well as the Vieweg Verlag, for giving us the opportunity to publish this book.

Kozo Fujii and George S. Dulikravich   August, 1998

# Contents

# Multi Objective Aerodynamic Optimisation by Means of Robust and Efficient Genetic Algorithm

Carlo Poloni
Dipartimento di Energetica
Università degli Studi di Trieste
Via Valerio 10, 34100 Trieste, ITALY
e-mail: poloni@univ.trieste.it

## Summary

In this paper the use of Genetic Algorithms for multi objective optimisation in aerodynamic optimisation is outlined. After a review of existing GA methodologies the operators considered at present the most promising one are described. A simple mathematical test is used for preliminary algorithmic perfomance while in more applicative cases the pressure reconstruction problem of two conflicting aerodynamic profiles is used as benchmark. A full potential transonic solver is at first used showing the performances of the optimisation algorithm employed while final results are obtained using a commercial Navier-Stokes solver with k-e turbulence modelling to reconstruct the geometry of two airfoils working at Mach=0.2 Re=5E6 and Mach=0.77 Re=19.6E6.
Even thogh the test case presented might not have a practical application, it shows that direct multi objective optimisation with Navier Stokes solver can be faced with GA.

## Introduction

In recent years much interest has been addressed to the use of Genetic Algorithms as general purpose optimisers and a large number of examples of engineering application can be found in the literature [1]. Recently this technique has been introduced even in the case of aerodynamic design where GA has been used for aerodynamic shape optimisation in the case of viscous incompressible flows [2] and in the case of inviscid compressible flows [3]. Even though several other example of GA application to aerodynamic problems can be found in the literature of more recent years [4], [5] and the optimisation of a 3D wing done using the parallel super computer and a full Navier Stokes code was recently presented [6] the main concern related to the use of Genetic Algorithm for aerodynamic design is the computational effort needed for the accurate evaluation of a design configuration that, in the case of a crude application of the technique, might leads to unacceptable computer time if compared with other more classical algorithms [7].

Three main issues makes however GA more than attractive and maybe unique among the aerodynamic design optimisation methods: GA are usually much more robust than gradient based algorithm and can tolerate even approximate or noisy design objectives evaluation, GA can be efficiently parallelised and can therefore take full advantages of the massively parallel computer architecture, GA can directly approach a multi objective optimisation problem [8], [9], [10].

Most real-life design procedures are complex tasks that have to deal with multi disciplinary environments, not always clearly defined targets, constraints to be satisfied. In this sense even tough the target of the optimisation could be expressed with a single expression like: "do the best possible design", the optimisation process must consider several different usually conflicting objectives and the compromise obtainable might not be a-priori known. The possibility of looking not only for a single good solution but for a set of solutions (the Pareto Set) [11] that satisfy different levels of compromise might be of great help to the decision maker that must select the most suitable one.

## Optimisation Algorithm

Typical optimisation problems are usually solved by means of "hill climbing procedure" possibly based on local gradients of a stated "cost function". The typical drawback of this approach is the fact that the search for improvements is done efficiently but is done locally.

On the other hand probabilistic optimisation techniques can be used to examine a large but discrete configuration space in order to find a "good solution" possibly close to the global optimum.

The basic idea of Genetic Algorithm is the one of a simplified biological evolution that adapt the specie to the environment through simple transformation of coded configurations.

As in the natural process of reproduction the genetic information stored in a chromosomal string of mating individuals are used to create new individual genetic code.

The design procedure can be viewed as an evolutionary process with the following terms correspondence:

| | | |
|---|---|---|
| gene | = | design parameter |
| individual | = | design configuration |
| generation | = | group of configurations |
| fitness function | = | design quality |
| social success | = | optimality of the design . |

Two major characteristics of a simulated biological evolution needs to be underlined:

- GA starts from a number of possible solutions and not from a single starting point as in the case of gradient based techniques
- the "fitness function" and the "social success" can be subjective leading to a natural codification of a multi objective optimisation process.

The key points of the GA are the operators used for selection and reproduction that highly influence the robustness and the efficiency of the algorithm.

In this paper some of the possible operators are briefly illustrated pointing out the most suitable to engineering problems, like aerodynamic optimisation, for which the objective functions topology is not too complex (not-deceptive) but the computation of the objectives is computationally intensive.

2

The GA algorithm has in the end the following structure:

```
do ng generation

    do nind individuals
        translate bits into variables
        compute objectives => interface to analysis
    end do

    do some statistics on the population individuals

    do Create a new population:

            by cross over:
                select individual
                and reproduce

            by mutation:
                select individuals
                and mutate

    end do
end do
```

Selection schemas

Roulette Wheel selection is the first used and most popular operator. A selection probability proportional to its fitness is assigned to each individual in the population. If no scaling, i.e. adimensionalisation of fitness in the *min max* range, is applied premature convergence occur that can be avoided only with large population size. The operator is robust but computationally expensive.

Tournament Selection overcome the problem of fitness scaling with direct comparison of fitness value: on a $t_{size}$ group of individuals the best is selected. This type of selection is generally considered more efficient and more robust than roulette wheel.

Local Geographic Selection [12], elsewhere named as step-stone island model, is a particular case of Tournament Selection. The *n-size* individuals participating to the tournament are not selected randomly in the population but through a local random walk in the neighbourhoods of a given individual being the population distributed in a N dimensional grid.

Cross-over operators

Two point cross-over is the most classical operator for reproduction and is one of the operators that offer the highest robustness to the search. Two point are randomly chosen and the genetic materials (the design variables) are exchanged between the parent variables vectors.

Directional cross-over is slightly different and assume that a "direction of improvement" can be detected comparing the fitness values of two reference individuals. In [13] a novel operator called Evolutionary Direction Crossover was introduced and it was shown that even in the case of a complex multimodal function this operator outperforms classical crossover. The direction of evolutionary improvement is evaluated by comparing the fitness in generation *j* of the individual *i* with the fitness of the two parents belonging to generation *j-1*.

The new individual is then created moving in a randomly weighted direction that lies within the ones individuated by the given individual and his parents. A similar concept can be however applied on the basis of directions not necessarily linked to the evolution but detected selecting two other individuals in the generation. The new schema for the Directional Crossover becomes:

1. for all individuals $i$
2. select individual $i1$, select individual $i2$
3. create the new individual as:
$$\bar{x} = \bar{x}_i + S \cdot sign(F_i - F_{i1}) \cdot (\bar{x}_i - \bar{x}_{i1}) + T \cdot sign(F_i - F_{i2}) \cdot (\bar{x}_i - \bar{x}_{i2}).$$

Where $S$ and $T$ are random numbers in the interval $S,T \in [0,1]]$, $F$ is the value of the fitness function for the corresponding vector of variables $\bar{x}$.

GA for Multi Objective Optimisation

The multi-objective optimisation problem can be expressed as follow:

$$\max F_i(\bar{x}) \quad for \quad i = 1,n$$
$$g_j \leq 0 \quad for \quad j = 1,m$$

and it is obvious that in general the solution is not unique if the functions are not linearly dependent.

With the introduction of the Pareto dominance concept it is possible to divide any group of solutions into two subgroups: the *dominated* and the *non dominated* one.

Solutions belongings to the second group are the "efficient" solutions, i.e. the ones for which it is not possible to increase any objective value without deteriorating the values of the remaining objectives.

In more formal terms and in the case of maximisation problems it is possible to say that the solution $\bar{x}$ dominates $\bar{y}$ if the following relation is true:

$$\bar{x} >_p \bar{y} \Leftrightarrow ( \forall i \quad F_i(\bar{x}) \geq F_i(\bar{y}) ) \cap ( \exists j \quad F_j(\bar{x}) > F_j(\bar{y}) ).$$

Classical optimisation algorithms are capable, under strict continuity and derivability hypothesis, of finding the optimal value only in the single objective case and therefore the problem of finding the group of non dominated solutions (the Pareto Set) is reduced to several single objective optimisation where the objective becomes a weighted combination of the objectives called utility function $U$:

$$U_k = \sum_{i=1}^{n} W_{ik} F_i(\bar{x}).$$

Where $\bar{x}$ is the vector of variables and $W_i$ are the weights for the objectives $F_i$.

While traditional optimisation algorithm do need the use of an utility function, the particular structure of GA can face the multi-objective optimisation problem in a more direct way developing populations in which the diversities follow the conflicting objectives.

Pareto-GA algorithms mainly differ from classical GA in the selection process even though other specific operators might be constructed.
A quick review of Pareto-GA techniques is given in the following.

Pareto Tournament Selection

As shown in [9], [10] the tournament selection can consider the Pareto concept as a basis for the tournament: the selected individual is the one that dominates the individuals taking part to the tournament. Most profitable implementation of this method are usually coupled with sharing.

Local Pareto Selection

An effective way of maintaining diversity in the population able to follow the conflicting objectives can be the use of Local Selection schema based on the Pareto dominance concept. In this case the population is placed on a toroidal grid and the members of the local tournament are chosen by means of a random walk in the neighbourhoods of the given grid point.

Ranking Based Selection (Non dominated Sorting GA)

With this approach shown in [14] and suggested in [15] the classical fitness function is substituted by a ranking value. To all the non dominated members a rank of one is assigned and then removed from the population. The non dominated members in the remaining part of the population are then assigned a rank of two and this process is continued until the entire population is ranked. To each individual is then assigned a fitness value based on its non domination ranking.

Local Geographic Selection based on Pareto Dominance

the dominant individual (in a multi objective sense) of a tournament defined by a local random walk is selected.

Local Tournament Directional Selection

Given a reference individual the tournament is made considering a linear combination of objectives with reference individual weightings.

Local Geographic Directional Selection

Given a reference individual the local geographic tournament is made considering a linear combination of objectives with reference individual weightings.

Sharing Function

Sharing function is the most widely used approach for diversity keeping. It consists in distributing available resources between similar competing individuals. The sharing is usually applied at genotype level or at phenotype level but in the case of Pareto-GA it might be more effective, as underlined in [14] to do it in the objectives space.

Vector Evaluated Genetic Algorithm

In this approach proposed by Schaffer and referred to as VEGA algorithm, the population of size $P$ is subdivided into $N$ sub-population each of them is addressing one of $N$ criteria. Typically as done in [8] the different criteria are expressed as different utility function with $N$ different weightings.

Multi Directional Crossover

When a multi objective optimisation task must be performed the fitness differences can be substituted with the differences of the respective scalar products of the finesses vector in the direction given by the fitness of the individual to be reproduced :

$$\bar{x} = \bar{x}_i + S \cdot sign(F - F1) \cdot (\bar{x}_i - \bar{x}_{i1}) + T \cdot sign(F - F2) \cdot (\bar{x}_i - \bar{x}_{i2})$$

where:     $F = \bar{F}_i \times \bar{F}_i$, $F1 = \bar{F}_{i1} \times \bar{F}_i$ and $F2 = \bar{F}_{i2} \times \bar{F}_i$.

The selection of individuals $i1$ and $i2$ can be done using any available selection schema.
In this way global information about the fitness landscape are retained in the whole population while each individual maintain and improve their specialisation through the evolution.


# A simple mathematical test

While the use of mathematical tests with growing complexity are widely employed for performance assessment of single objective optimisation algorithms, in the field of multi objective optimisation the subject is much less defined. It has been therefore decided to simply combine two mathematical functions defined in the same variables interval and use them as a multi objective test.
Let be:

$$x, y \in [-\pi, \pi]$$
$$F_1(x, y) = -[1 + (A_1 + B_1)^2 + (A_2 + B_2)^2]$$
$$F_2(x, y) = -[(x + 3)^2 + (y + 1)^2]$$

where:

$$A_i = \sum_{j=1}^{2} (a_{i,j} \cdot sin(\alpha_j) + b_{i,j} \cdot cos(\alpha_j))$$

$$B_i = \sum_{j=1}^{2} (a_{i,j} \cdot sin(\beta_j) + b_{i,j} \cdot cos(\beta_j))$$

$$a = \begin{bmatrix} 0.5 & 1.0 \\ 1.5 & 2.0 \end{bmatrix} \quad b = \begin{bmatrix} -2.0 & -1.5 \\ -1.0 & -0.5 \end{bmatrix} \quad \alpha = \begin{bmatrix} 1.0 & 2.0 \end{bmatrix}.$$

Two functions $F_1$ and $F_2$ of the two variables $x$ and $y$ to be *max* in the given interval. The first is the function used in [16] and the second is a paraboloid shape with its maximum in [-3,-1]. Figure 1 shows the contour plot of a linear combination of the two objective function given as: $F = k \cdot F1 + (1 - k) \cdot F2$ where $0 \le k \le 1$ while figure 2 shows the results in the objectives plane of an exhaustive calculation in a $64 \times 64$ grid.

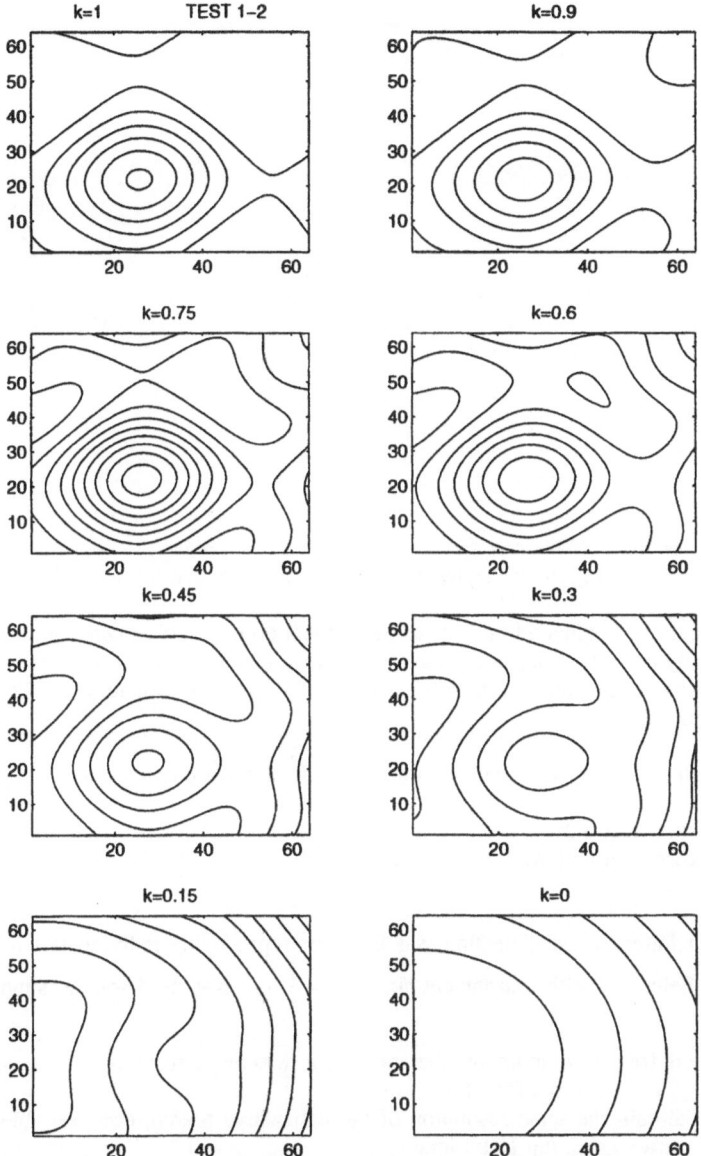

Figure 1: Contour of linear combination of functions *F1* and *F2*

It can be noted that the Pareto front, underlined using circles, is divided into two region that belongs to different portion of the variables space with the consequence that even in such a

7

simple case the linear combination of only two objectives into one can produce a non trivial Pareto front that can not be obtained by subsequent single objective minimisation done relaxing the linear combination parameter $k$.

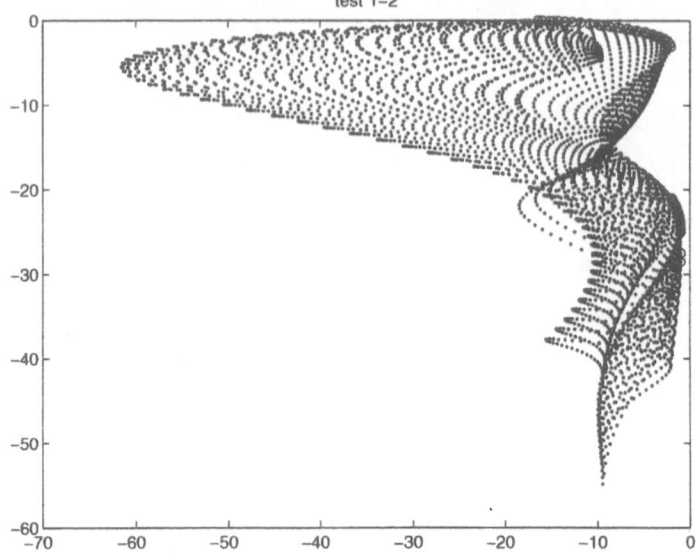

Figure 2: The test functions in the objective space

Because in real life engineering problems the evaluation of a design configuration might be computationally expensive, it has been decided to test the Pareto GA with a strict limit on the number of calls to the fitness function calculation routine corresponding to 32 individuals times 8 generation.

Three different runs have been made in order to compare three possible cases:

1. classical cross-over only
2. classical cross-over and MDC for 50% of the cases
3. MDC only .

In all cases a Pareto Local Selection was used for reproducing individuals while a Pareto Tournament Selection with tournament size $t_{size} = \frac{1}{5} p_{size}$ was used for the simply selected individuals.

The same seed for the random number generator was used resulting in the same initial population randomly selected for all the runs.

In order to maintain the same resolution of the exhaustive search, each variable uses 5 bit leading to an individual string 10 bit long.

The case that uses the Multi Directional Crossover in the 50% of the cases (PCD=0.5) outperforms both the algorithms with classical crossover or MDC alone. An intuitive explanation of this is the fact that the MDC acts more as a stochastic hill climber being more efficient in exploitation than in exploration.

The number of points on the Pareto curve detected reflects this characteristic too: case1 has 13 points, case2 has 18 points and case3 has 16 points.

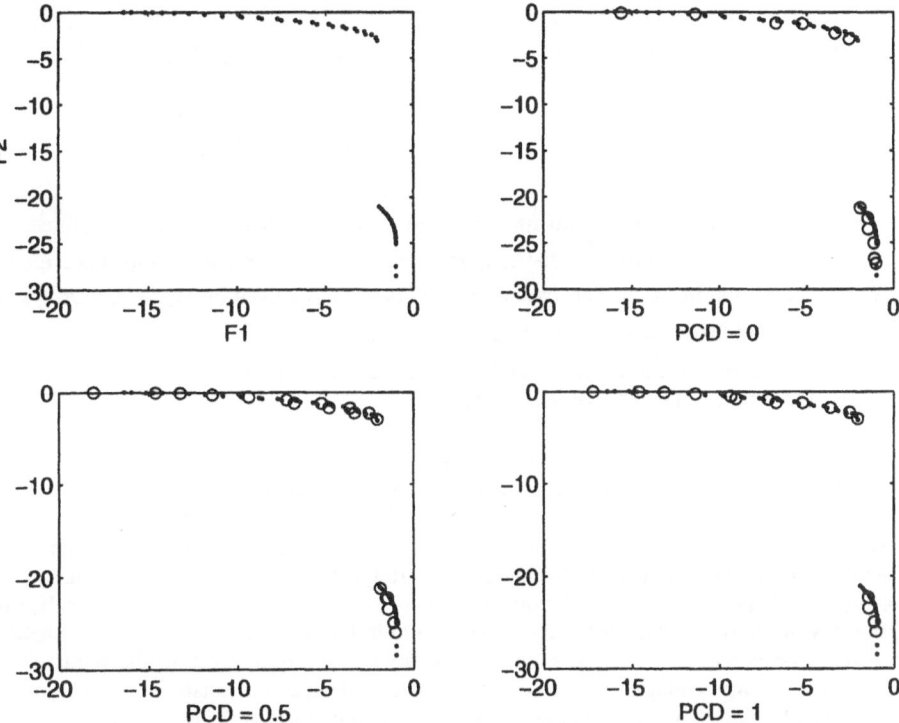

Figure 3: Comparison of the Pareto front obtained for the three strategies adopted

## Pressure reconstruction of an airfoil

The optimisation problem that is here addressed deals with the construction of a compromise airfoil that should withstand with two incompatible working condition: high lift in subsonic flight and low drag in transonic flight. The test case considered is a modified version of the test case shown in [17]. The contour geometry of two existing airfoils one for high lift and the other for low drag are given in a tabulated form. The pressure distribution is therefore to be computed by the same flow solver that is used for the following pressure reconstruction. No assumption is made on the flow solver to be used and therefore it has been chosen to use a full potential transonic code to execute some preliminary calculation in order to verify the performances of the optimisation algorithm while, with the tuned parameters, a Navier-Stokes with ke turbulence modelling is then used.

The two design point at which a specific profile is given are:

| airfoil | case | $\alpha_{inc.}$ (°) | Mach | Re |
|---------|------|---------------------|------|-----|
| high lift | 1 | 10.8 | 0.2 | 5.0E6 |
| low drag | 2 | 1.0 | 0.77 | 19.6E6 |

Once the reference pressure distribution are computed, the two-point design problem is defined as the maximisation of the two objective functions that in the inviscid case are:

$$F_1 = -\int_0^1 \left(C_p^1(s) - C_{ptar}^1(s)\right)^2 ds \quad F_1 = -\int_0^1 \left(P_1(s) - P_{1tar}(s)\right)^2 ds$$

$$F_2 = -\int_0^1 \left(C_p^2(s) - C_{ptar}^2(s)\right)^2 ds \quad F_2 = -\int_0^1 \left(P_2(s) - P_{2tar}(s)\right)^2 ds$$

where $s$ is the fractional arc length measured along the airfoil contour to be defined, $C_p^i$ represents the pressure coefficient distribution in the inviscid case and $P_i$ the relative static pressure distribution in the viscous case at design condition $i$ (given by $\alpha_i$, Mach, Re) while $C_{ptar}^i$ and $P_{itar}$ are the target quantities at design condition.

It must be noted that the minus in front of the integrals is used in order to transform the usual minimisation problem into a maximisation one.

## Airfoil shape parametrisation

Any optimisation approach needs a parametric representation of the problem and this task is usually matter of studies by itself. In the case of aerodynamic design by means of direct flow solvers one has always to cope with computationally expensive objective function evaluation. Efficiency of the solvers and of the optimisers might in fact vanish if an efficient geometrical representation is not adopted. The number of degrees of freedom given to the parametrised shape must be large enough to explore a wide range of shapes but small enough in order to limit as much as possible the number of variables involved in the problem.

Satisfactory results have been obtained using an assembly of Bezier curves [18] for the definition of parametrised shapes and therefore even in the case of the airfoil a similar approach have been used. A Bezier polynomial of order $n$ is defined by:

$$b(t) = \sum_{i=0}^{n} P_i \Phi_i(t) \quad \Phi_i(t) = \left(\frac{n!}{i!(n-1)!}\right)$$

being $P_i$'s the vertices of the Bezier control polygon.

In the example here shown 18 parameters define the airfoil shape, four of them determines the camber line defined by means of one cubic Bezier curve while the remaining 12 determines the location of the control points related to the upper and lower side of the thickness distribution that is superimposed to the camber line. Each side of the thickness distribution is given by two Bezier polynomials one of order three for the leading edge and one of order four for the rear part.

A schematic representation of the variables' meaning is given in the figure 4. While in figure 5 the profiles obtainable with the minimum and maximum values allowed for the 18 parameters. It can be seen that the allowed variation range is rather large with shapes that, if used as starting point of a hill-climbing algorithm, would not bring to a correct solution of the optimisation problem [19].

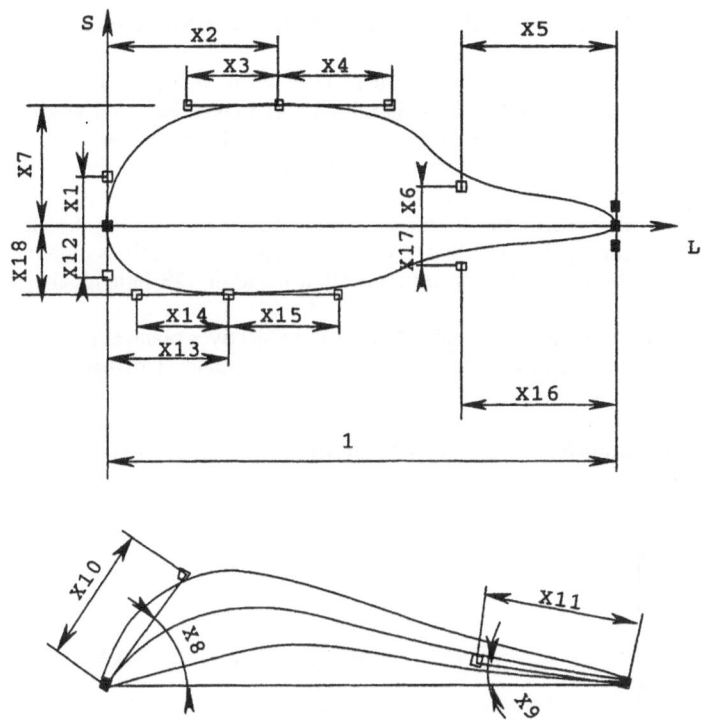

Figure 4: Airfoil parametrisation

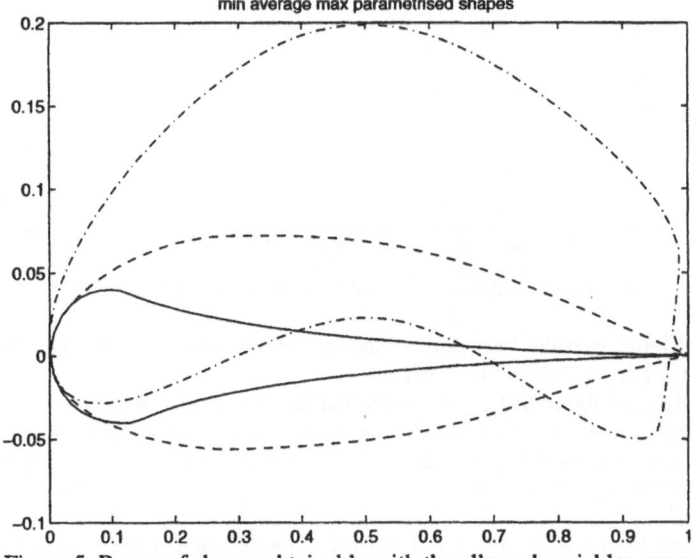

Figure 5: Range of shapes obtainable with the allowed variables range

## Results for the airfoil pressure reconstruction - inviscid case -

The problem is the same as the one faced in [19] where it was demonstrated that it was possible to reconstruct the airfoil shape in all the given conditions and where it was shown that a hybrid approach using traditional optimisation algorithms in conjunction with GA can provide the best results. This work focuses instead on the Pareto GA approach and therefore only the multi objective case is here considered.

For this problem three different run are compared, once more varying the amount of MDC used but also considering the evolution of the Pareto front as the optimisation proceeds.

All the tests consider the same starting population made of 128 individuals evolved for 64 generations.

Figure 6 shows the non dominated individuals in the objectives plane at generation 64 and the results for the three different values of PCD (Directional Crossover Probability) are compared.

No major differences are visible even though when only MDC is used several points in the Pareto front are dominated by the solutions obtained with classical crossover only (PCD=0) or with a mixed approach (PCD=0.5).

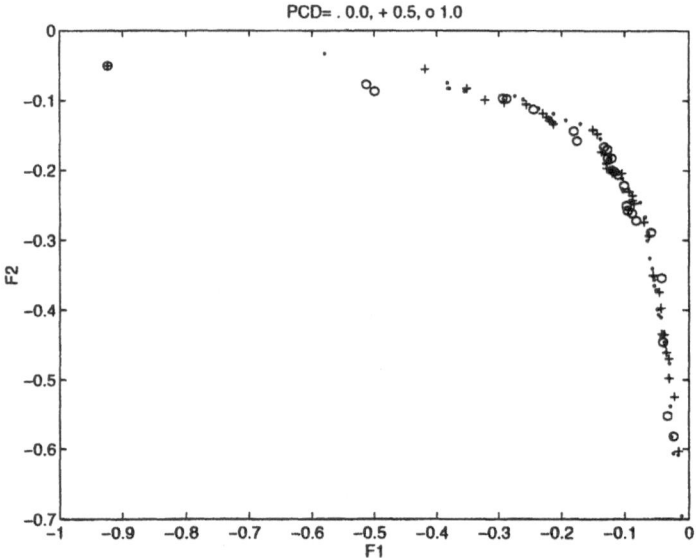

Figure 6: Pareto set detected by the three tested strategies

The strong difference in the behaviour of the algorithms with MDC becomes evident if the evolution of the Pareto front is examined.

From the analysis of figure 7 it can be noted that the main effect of the directional crossover is a much faster *convergence* that produce almost the same set of efficient solutions as the final one already from the 4th to 8th generation saving in this way one order of magnitude of cpu time for almost the same result.

It is clear that classical crossover can maintain better diversity in the population and therefore obtain a little better resolution of the Pareto front but with much higher computational cost.

12

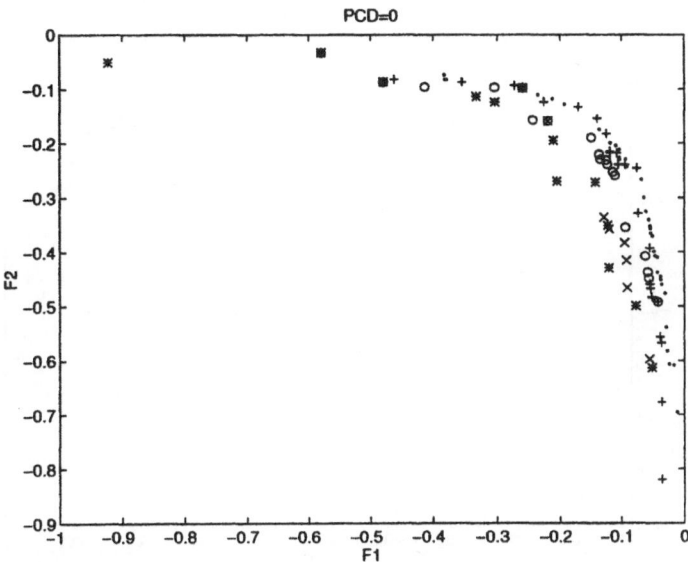

Figure 7b: Pareto front evolution at generation: *4, +8 x16 o32 .64

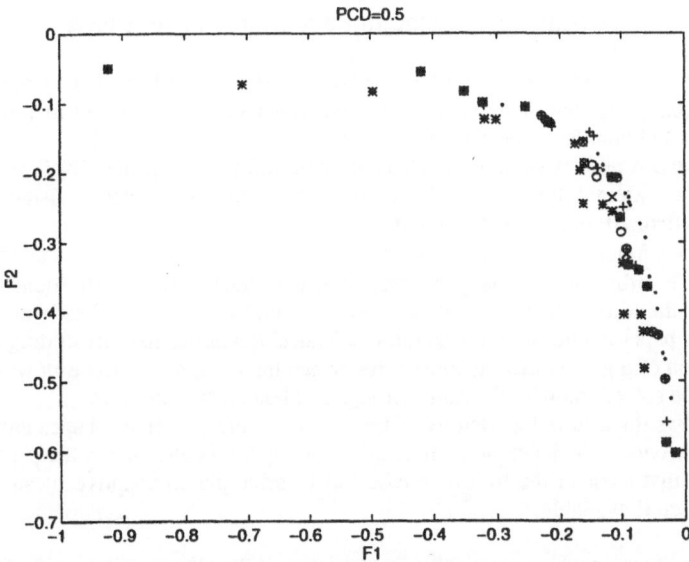

Figure 7b: Pareto front evolution at generation: *4, +8 x16 o32 .64

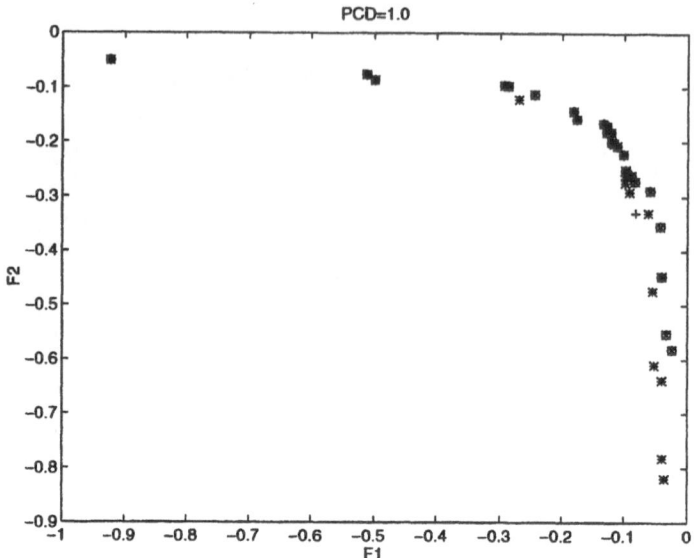

Figure 7c: Pareto front evolution at generation: *4, +8 x16 o32 .64

## Pressure reconstruction problem - viscous case -

In the viscous case the commercial CFD solver STAR-CD [20] is used to model the flow while a Voronoy-Delaune automatic mesh generator recently developed in house [21] was used for grid and boundary condition generation.

The typical mesh used is of about 3500 cells with 128 points on the airfoil surface and 64 points in the external boundary. The convergence criteria adopted considered a global residual less than $10^3$ or 600 as maximum number of iteration.

The turbulence model adopted is a standard $\kappa$-$\varepsilon$ with wall function with boundary condition equal to 10% turbulence intensity at inlet with a characteristic length equal to the mean thickness of the airfoil. It must be noted that no tuning of the solver has been made as the focus of this paper is not the optimisation of physical quantities like lift to drag ratio but the reconstruction of a given airfoil geometry for which the pressure distribution was determined with the same solver and with the same settings and boundary conditions.

Figure 8 shows the grid in the vicinity of the airfoil for the two given shapes and the relative pressure distribution. As it can be seen no grid refining has been made in the proximity of the shock wave that exist in the transonic case but in principle an adaptive mesh solver could have been used if available.

Two optimisation run where made: the first one as a multi objective case with 64 individuals for 8 generations and the second one with 32 individuals for 4 generations as a single objective case in order to improve the quality of the high lift profile reconstruction.

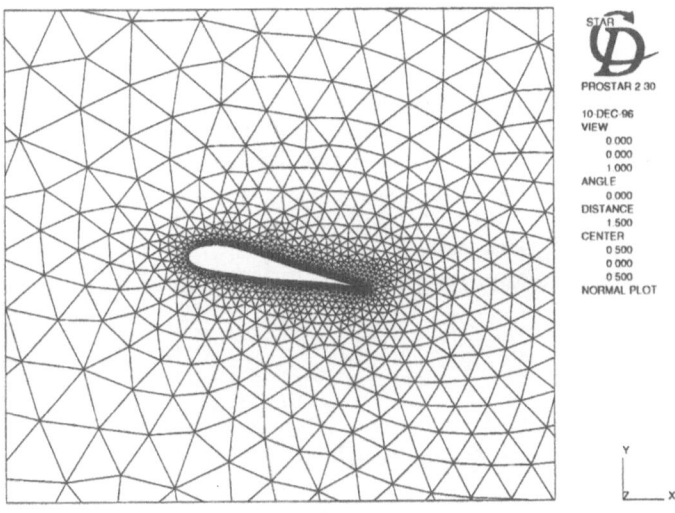

Figure 8a: High Lift profile, grid used

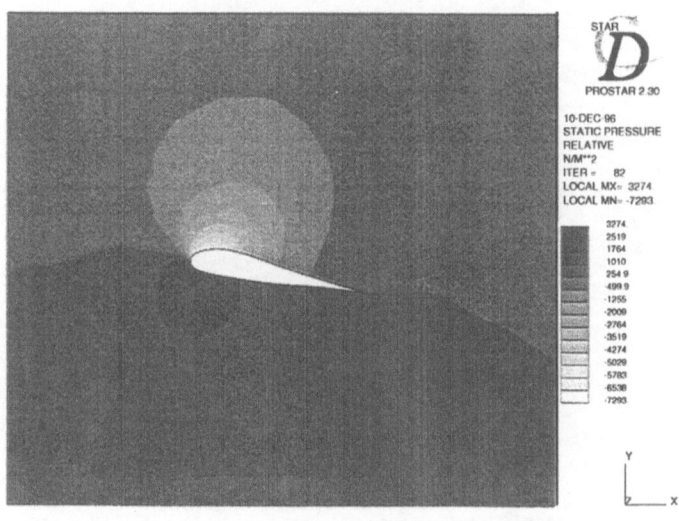

Figure 8b: High Lift profile pressure distribution
Mach=0.2 Re=5.0E6

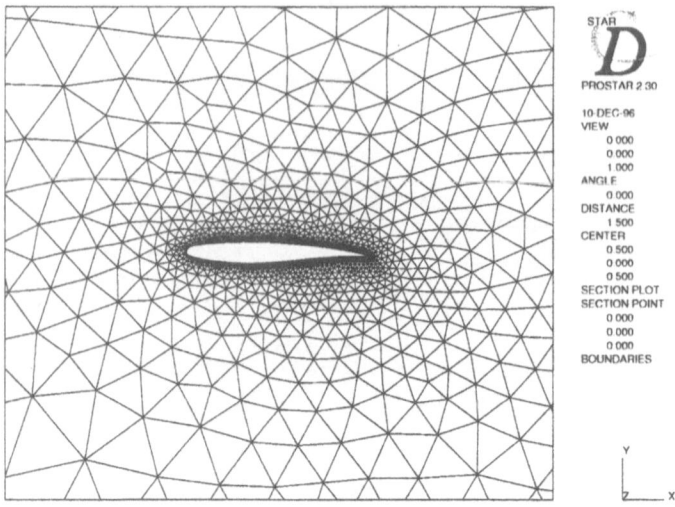

Figure 8c: Low Drag profile grid used

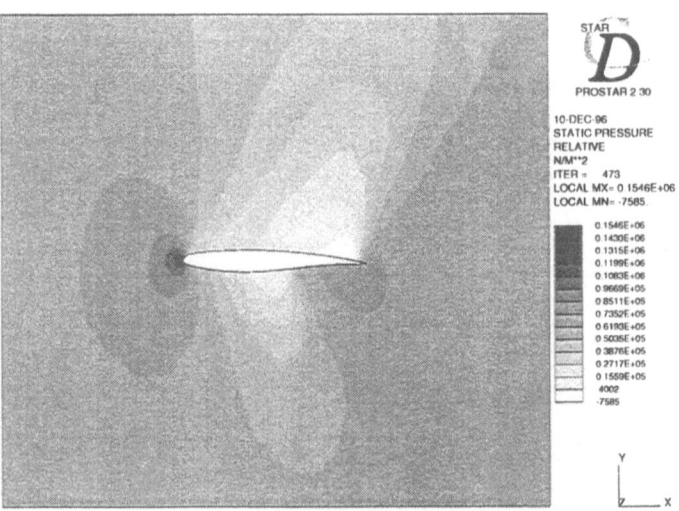

Figure 8d: Low Drag profile pressure distribution
Mach=0.77 Re=19.6E6

In figure 9 the RMS of the difference in pressure between the given profile geometries and the computed one is reported pointing out the existence of several non dominated solutions. that represent possible compromise solution to the pressure distribution given by the two different profiles.

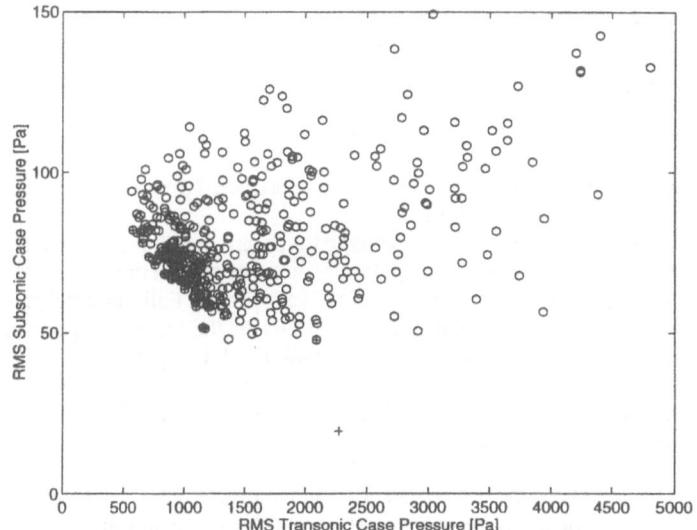

**Figure 9: The Pareto set detected in the objective space**

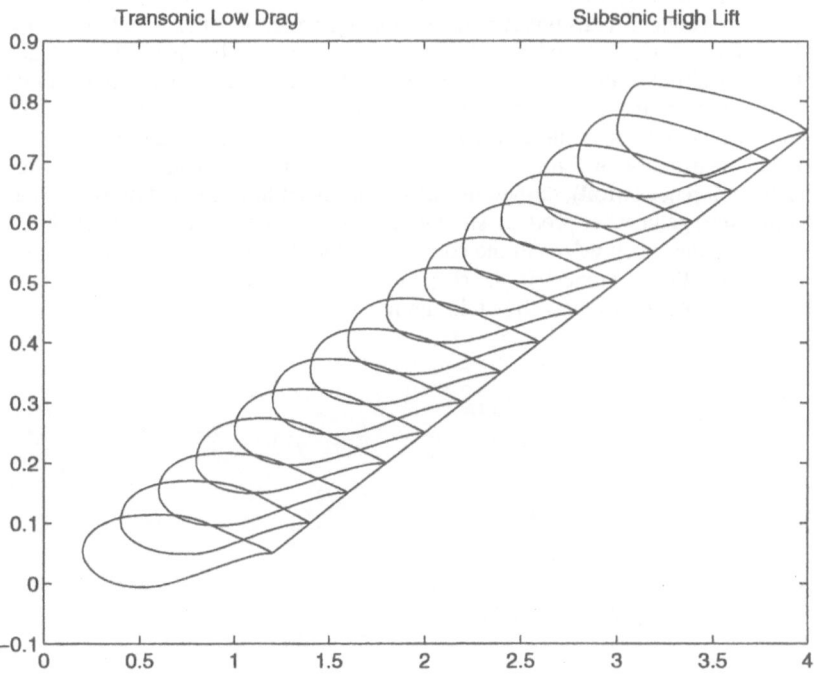

Transonic Low Drag　　　　　　　　　　　Subsonic High Lift

**Figure 10: shapes of the profiles in the detected Pareto set**

The geometry of the 15 airfoils obtained with the multi objective pressure reconstruction is reported in figure 10 where a clear transition from high lift shapes to low drag ones can be seen.

Comparison between the reconstructed geometry inviscid/viscous case.

A quantitative comparison of the quality of the reconstructed geometries can only be made for the two extrema of the Pareto frontier.
Figure 11a and figure 11b shows the reconstructed pressure distribution obtained for the two cases considered. The behaviour of the pressure distribution is correctly captured in both cases and both with the Navier Stokes solver as well as with the Full Potential solver.
The major differences are to be found in the proximity of the leading edge of the transonic airfoil for which the given geometry have a small local bump that cannot be correctly modelled with the chosen airfoil parametrisation. The transonic airfoil geometry present a strong shock wave in the upper side as well as a weak shock in the lower side that are both correctly located in the reconstructed geometry.

Bigger differences, mainly in the transonic case, can be noted if geometries are compared.
While the trailing edge of the transonic airfoil that in the end determine the position of the shock wave is practically the same for the full potential calculations, a clear difference exist in the case of the reconstructed geometry obtained with the Navier Stokes calculation.
Even though further investigation on this topic should be made this fact might introduce the suspect that the Navier Stokes pressure reconstruction problem with k-e turbulence modelling could have non-unique solution due to the presence of the boundary layer that directly influence the position of the rear edge shock wave. GA in this case might have found one of the possible solutions that satisfactorily reproduce the given pressure distribution. This fact is even more evident looking at the correlation between figures 11a. 12a and 13.
The pressure distribution of figure 11a are the one of the respective airfoil geometries having full line (the reference airfoil), dashed line and dash-dotted line reported in figure 12a.
Slightly different geometries produce similar pressure distribution if the Navier-Stokes solver is used locating the shock waves in the correct position. It is not the same if an inviscid code is used for which the differences in the shape of the upper side of the airfoil presents a shock in a slightly different position as shown in figure 13.

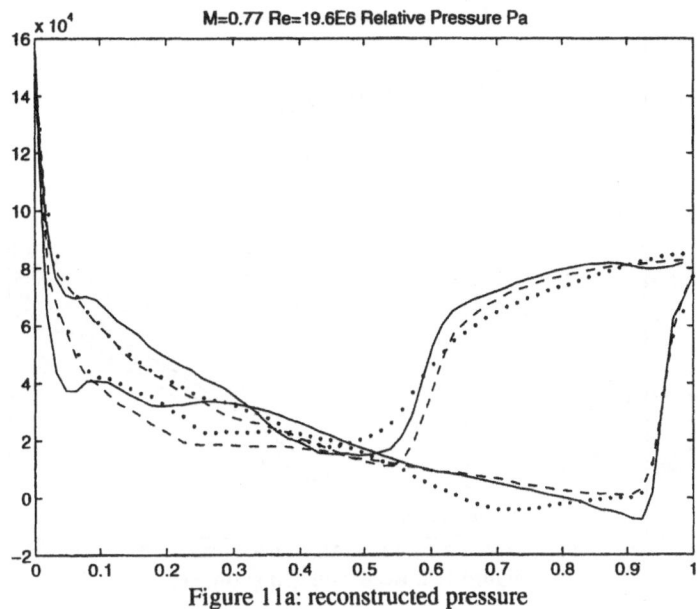

Figure 11a: reconstructed pressure

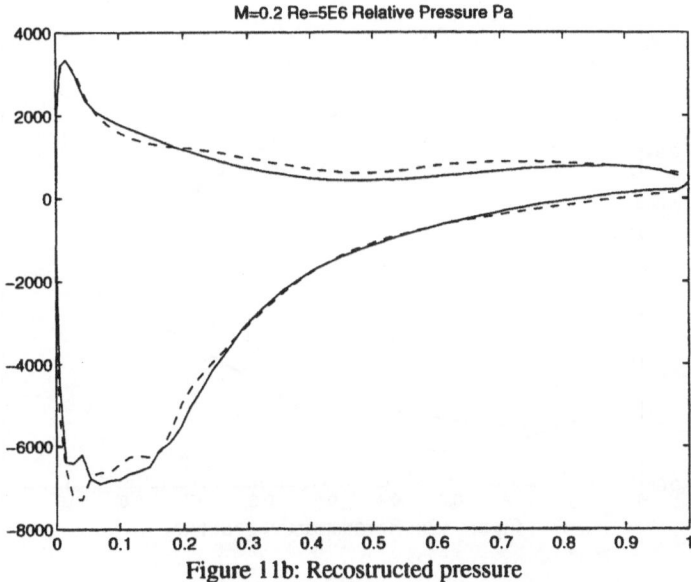

Figure 11b: Recostructed pressure

19

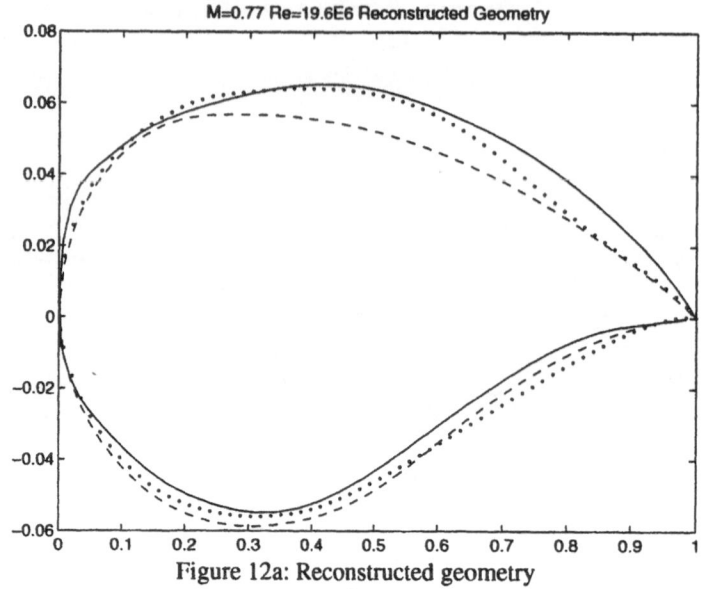

Figure 12a: Reconstructed geometry

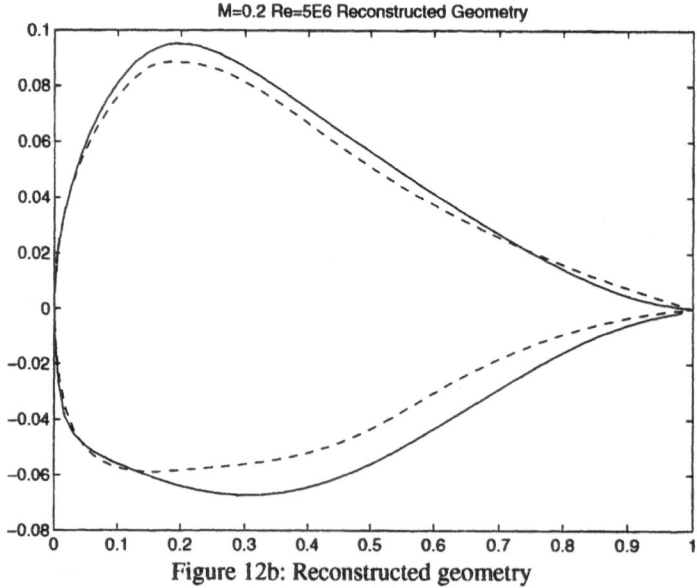

Figure 12b: Reconstructed geometry

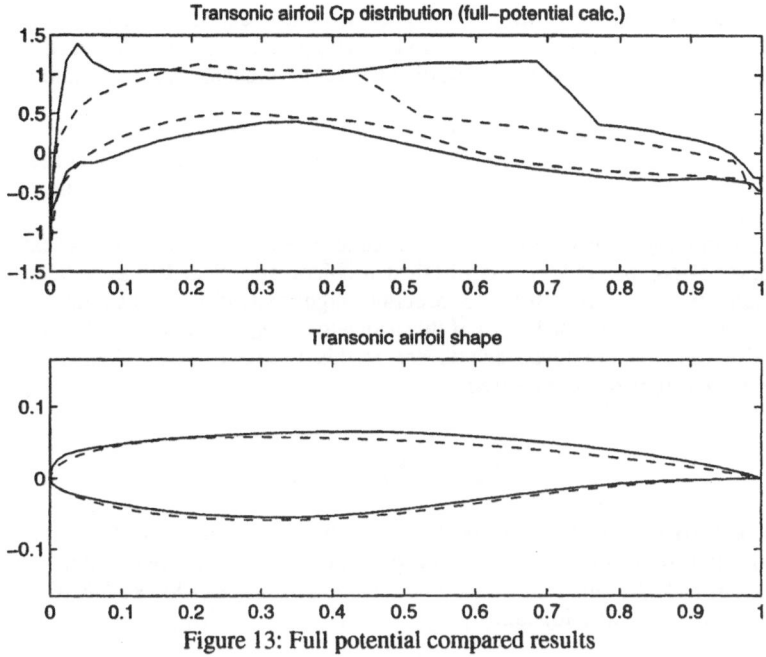

Figure 13: Full potential compared results

## Computational results

All the calculation done for this paper where made on a Sun Ultra 1/140 with 64 Mbytes RAM. While a single full potential simulation takes approximately 1s a Nervier Stokes calculation takes 90s for the incompressible subsonic case and 500s in the transonic case.

With these figures the NS Pressure reconstruction problem took about 4 days plus 3.5 hours for the refinement of the subsonic profile reconstruction.

It must be noted that the Work Station used is not a high-end machine and still it was possible to run a 2D NS optimisation in reasonable time.

Elsewhere [22] data published about computational performances of a parallel version of a GA based optimisation code have shown that 98% parallel efficiency on a 64 processor machine can be obtained with a linear speed up of the computational process and therefore more conplex problems can be considered tractable on high performance parallel computers.

## Conclusion

In this paper after a quick review of existing technologies for multi objective optimisation with Genetic Algorithm some specific operators are illustrated showing that the computational cost of the optimiser can be highly reduced and made comparable with the one of other optimisation tools still maintaining most of the robustness of evolutionary optimisation.

The described algorithm has then been applied to a two objectives aerodynamic problem handling the case of pressure reconstruction of compromise airfoils that have to work in conflicting operating condition like subsonic/high angle of attack and transonic/low angle of attack.

Both inviscid and viscous case where run using a full potential solver in the first case and a Navier Stokes with keg turbulence modelling commercial solver in the second case.

First of all the results obtained have shown that a 2D viscous case can be handled in a small-size workstation in a reasonable time.

From the preliminary results obtained it came out also that the optimisation process can make evident physical complexities of the physical problem modelled. It was in fact clear that multiple solution to the geometry reconstruction might exist if a viscous code is employed. Further investigation are needed to see if this is due to the presence of several local minima in the objective function employed and to see if this is due to physical phenomena or to weakness of the turbulence model used.

## Acknowledgements

The help of B.Niceno in the implementation of the grid generator and the many fruitful discussion with Prof. Mosetti have been indispensable for the completion of this work.
This work was partially financed by EC under the Esprit project FRONTIER n° 210082 and partially with MURST 40% founding.

## References

[1] Goldberg, D.E., Kelsey M. and Tidd, C.: Genetic Algorithm: A Bibliography, (IlliGAL Report n. 92208), University of Illinois atUrbana-Champaign, Illinois Genetic Algorithm Laboratory, Urbana, USA (1992).

[2] Mosetti, G, Poloni, C.: Aerodynamic Shape Optimization by Means of a Genetic Algorithm, 5th International Symposium on Computational Fluid Dynamics, Sendai, Japan (1993).

[3] Quagliarella, D. and Della Cioppa, A.: Genetic Algorithms applied to the Aerodynamic Design of Transonic Airfoils, 12th AIAA Applied Aerodynamics Conference, Colorado Springs, CO USA, AIAA-94-1896, (1994).

[4] J.Periaux, G.Winter, M. Galan, P. Questa ed.: Genetic algorithms in engineering and computer science, John Wiley & Sons, England, (1995).

[5] AIAA-ASME Joint Conferece on Multidisciplinary Design and Optimisation (1996).

[6] Obayashi, S. and Oyama A. : Three-Dimensional Aerodynamic Optimisation with Genetic Algorithm, ECCOMAS 96 proceedings, Paris 9-13 September (1996).

[7] AGARD Report : Optimum Design Methods in Aerodynamics, AGARD-R-803, (1994).

[8] Hajela,P., Genetic search strategies in multicriterion optimal design, Structural Optimisation, Vol.4,pp. 99-107, Springer Verlag (1992).

[9] Horn, J. and Nafpliotis, N., Multiobjective Optimisation Using the Niched Pareto Genetic Algorithm, IlliGAL Report n. 93005, Illinois Genetic Algorithm Laboratory, Urbana, USA, (1993).

[10] Belegundu, A. et Al., Multiobjective Optimisation of Laminated Ceramic Composites Using Genetic Algorithms V}, AIAA paper 94-4363-CP, (1994).

[11] Eschenauer,H., Koski, J. and Osyczka, A. , Multicriteria Design Optimisation Procedures and Applications, Springer-Verlag, Berlin, (1990).

[12] Collins, R.J., Jefferson, D.R., Selection in Massively Parallel Genetic Algorithms, Proceedings of The Fourth International Conference on Genetic Algorithms, Morgan Kaufman Pub. Inc., San Diego, USA (1991).

[13] Yamamoto,K. and Inoue,O., New evolutionary direction operator for genetic algorithms, AIAA Journal, vol.33-10, pp.1990-1993, (1995).

[14] Michielssen, E. and Weile,D.S., Electromagnetic system design using genetic algorithms, Genetic algorithms in engineering and computer science, pp. 345-369, John Wiley & Sons, England, 1995.

[15] Goldberg, D.E., Genetic Algorithms in Search, Optimisation and Machine Learning, Addison-Wesley, Reading Mass, USA, (1988).

[16] Van der Velden, A., Tools for applied engineering optimisation, AGARD-R-803, Optimum Design Methods for Aerodynamics, Special Course at the VKI, Rhode-Saint-Genese, Belgium, (1994).

[17] Labrujere, Th.E., Residual correction methods for aerodynamic design, AGARD-R-803, Optimum Design Methods for Aerodynamics, Special Course at the VKI, Rhode-Saint-Genese, Belgium, (1994).

[18] Arcilla, A.S., J.Hauser,J., Eiseman,P.R.,Thomson, J.F., editors: Numerical Grid Generation in Computational Fluid Dynamics, North Holland, (1991).

[19] Poloni, C.; Hybrid Genetic Algorithm for Multiobjective Aerodynamic Optimisation, Genetic algorithms in engineering and computer science, pp. 397-415, John Wiley & Sons, England, 1995.

[20] Star-CD User Manuals, Computational Dynamics Ltd. (1996).

[21] Niceno, B. : EASY numerical grid generator (soon available on http://.............) .

[22] Poloni,C. , Ng, D., Fearon, M. : Parallelisation of Genetic Algorithm for Aerodynamic Design Optimisation, Adaptive Computing in Engineering Design and Control, Plymouth University, UK, March (1996).

# Inverse Optimization Method for Aerodynamic Shape Design

Shigeru Obayashi

Department of Aeronautics and Space Engineering
Tohoku University
Sendai, 980-77, JAPAN
e-mail: obayashi@ad.mech.tohoku.ac.jp

## Summary

Characteristics of aerodynamic optimization have been discussed through wing shape design problems. It has been demonstrated that distribution of the objective function can be extremely rough even in a simplified problem. In such a situation, Genetic Algorithm (GA) is more effective than a simple hill-climbing strategy. GA, however, requires a large amount of computational time. To alleviate this, the inverse optimization method is developed and discussed in this paper. This method optimizes target pressure distributions for the inverse problem. In the two dimensions, viscous drag is minimized under specified lift and airfoil thickness. In the three dimensions, multi-objective GA (MOGA) based on the Pareto ranking method is employed. The present design procedure allows the minimization of the induced drag, while maintaining the straight isobar pattern of pressures. The resulting procedure is shown to be successful when applied to transonic wing design.

## Introduction

Aerodynamic numerical optimization methods are categorized into two classes[1]: direct and inverse numerical optimization methods. The direct numerical optimization methods are formed by coupling aerodynamic analysis methods with numerical optimization algorithms. They minimize (or maximize) a given aerodynamic objective function by iterating directly on the geometry. The geometry is represented either by a general function, such as polynomial and cubic splines, by a linear combination of known airfoils, or by a basic shape plus a combination of typical geometry perturbations.

In this chapter, the direct approach to simplified airfoil shape optimization is first considered and performance of the existing optimization algorithms is evaluated. One of them is the gradient-based method (GM). Another is the Simulated Annealing (SA) [2]. SA is a heuristic strategy for obtaining near-optimal solutions. SA derives its name from an analogy to the annealing of solids. The other is GA, one of the evolutionary algorithms.

Among the numerical optimization algorithms, gradient-based methods have been the most widely used. The optimum obtained from these methods will be a global optimum, if the objective and constraints are differentiable and convex [3]. In practice, however, it is very difficult to prove differentiability and convexness. One could only hope for a local optimum neighboring the initial point, provided the gradient is well-defined. Therefore, to determine the global optimum, one must optimize from a number of initial points and check for consistency in the optima obtained. In this sense, the gradient-based methods are not robust.

Evolutionary algorithms, Genetic Algorithms (GAs) in particular, are known to be robust [4] and have been enjoying increasing popularity in the field of numerical optimization in recent years. GAs are search algorithms based on the mechanics of natural selection and natural genetics. One of the key features of GAs is that they search from a population of points and not from a single point. In addition, they use objective function information (fitness value) instead of derivatives or other auxiliary knowledge. These features make GAs robust and thus attractive to practical engineering applications. GAs have been applied to aeronautical problems in several ways, including parametric and conceptual design of aircraft [5,6], preliminary design of turbines [7], topological design of nonplanar wings [8] and aerodynamic optimization using Computational Fluid Dynamics (CFD) [9-15]. The present result of the simplified aerodynamic optimization problem demonstrates the superiority of GA for aerodynamic optimization over the others.

Applicability of GAs to CFD problems, however, is limiting when a direct numerical optimization is used. CFD itself requires a large amount of computational time. The direct optimization approach requires CFD evaluation of each member of the population at every generation in GAs. As a result, a tremendous amount of computational time is required. The inverse approach therefore alleviates the computational time for engineering purposes. The main topic of this chapter is the development of the inverse optimization method using GAs.

The inverse numerical optimization methods deal with pressure distributions rather than geometry, to minimize, for example, drag under given lift and pitching moment. Since pressure is the primary force acting on an aircraft, one can design for desired aerodynamic characteristics by specifying pressure distributions. Once the target pressure distribution is optimized, the corresponding airfoil geometry can be determined by the inverse methods.

The inverse methods themselves form a class of powerful design tools. They solve the classical inverse problem of determining the aerodynamic shape that will produce the given pressure distributions. However, they leave the user with the problem of translating his design goals into properly defined pressure distributions that produce desired aerodynamic characteristics [16]. Although highly skilled designers are capable of producing successful designs, designers' efficiency can be improved by providing them with tools for target pressure specification. For this purpose, numerical optimization of target pressure distributions has been studied in [17,18]. This approach avoids most of the limitations of the standard inverse methods while requiring considerably less computational effort than the direct numerical optimization approach.

In the remaining part of this chapter, inverse optimization of wing shape is considered. The design of a wing usually proceeds in two steps. First, an airfoil section is designed. This reduces the three-dimensional design problem into a two-dimensional one. In [12], GA has been applied to optimize target pressure distributions around airfoils for inverse design methods. Pressure distributions are parameterized by B-spline polygons and airfoil drag is minimized under constraints on lift, airfoil thickness and other design criteria. Once the target pressure distribution is obtained, the corresponding airfoil geometry can be computed by an inverse design code by Takanashi [19] coupled with a Navier-Stokes solver [20].

Once the airfoil shape is designed, the next step in the wing design process is to determine the variation of the designed airfoil in the spanwise direction. The design principles for this step are essentially twofold. One is to preserve the two-dimensional performance as much as possible [21]. This is easily achieved by the inverse method by specifying the same chordwise pressure distribution along the wing span. The resulting wing has straight isobar pattern of pressure contours on the wing surface. The other is to minimize the induced drag essential to the three-dimensional wing. Incompressible flow theory predicts that the minimum induced drag is achieved by an elliptical lift distribution [22]. Therefore, the elliptical lift distribution is the key design criterion for the three-dimensional wing shape optimization.

Two design principles described above, however, contradict each other in general. Since the sectional lift is given by the chordwise pressure distribution, the elliptical lift distribution can be materialized by specifying the same chordwise pressure distribution along the wing span only if the wing has an elliptic planform. Tapered wings, however, have been adopted on the majority of aircrafts nowadays since they offer a compromised solution on account of their low induced drag, high maximum lift, low structural weight, good stowing provisions for the undercarriage and reasonable manufacturing cost [21]. As a result, the straight isobar pattern has been the primary design principle for transonic wings.

For better aerodynamic performance, three-dimensional target pressure distributions should be optimized in the chordwise direction to minimize the two-dimensional drag as well as in the spanwise direction to minimize the induced drag. Furthermore, the straight isobar pattern should be materialized to maintain the optimized two-dimensional performance. These requirements call for the multiobjective optimization of pressure distributions.

Multiobjective optimization (MO) seeks to optimize the components of a vector-valued objective function. Unlike the single objective optimization, the solution to this problem is not a single point, but a family of points known as the Pareto-optimal set [23]. Each point in this set is optimal in the sense that no improvement can be achieved in one objective component that doesn't lead to degradation in at least one of the remaining components.

Pareto-optimal solutions might be obtained by solving appropriately formulated single-objective optimization problems on a one-by-one basis, using methods, such as the weighted sum approach and the ε-constraint method. In contrast, by maintaining a population of solutions, GAs can search for many Pareto-optimal solutions in parallel [4]. In general GAs are not considered efficient since they require a large number of function evaluations. However, if GAs can sample solutions uniformly from the Pareto-optimal set, then they will turn out to be very efficient. Since GAs are inherently robust, the combination of efficiency and robustness make them very attractive for solving MO problems. Several approaches have been proposed and are called Multiple Objective Genetic Algorithms (MOGAs) [15,23-27].

The Pareto-based ranking method, as proposed by Fonseca and Fleming [25], is adapted to the present MOGA, since Pareto-based ranking and fitness sharing are keys to identifying a Pareto front. The techniques will be briefly explained in the following as well as the standard GA operators. Then the transonic airfoil design result will be shown. Finally, optimization of a typical transonic wing will be performed.

## Direct Optimization of Airfoil

1. Approximation concept

The airfoil design is to determine the contour $y$ for both upper and lower surfaces at every chordwise location $x$. How the $y$ coordinates are expressed determines the choice of design variables. Following [3], let's store airfoil designs in vector $Y^1$, $Y^2$, ..., $Y^N$, where these vectors contain the coordinates of the upper surface followed by those of the lower surface at given chordwise locations. They correspond to the existing airfoil shapes as the basis vectors. Thus an airfoil shape is defined as

$$Y = a_1 Y^1 + a_2 Y^2 + \cdots + a_N Y^N. \tag{1}$$

The design variables are now $a_1$ through $a_N$. This greatly reduces the number of design variables, say, in contrast with having the pointwise values of the $y$ coordinates over 50 chordwise locations. In the following, four basis airfoil shapes are used ($N = 4$). As defined in [3], $Y^1$, $Y^2$, $Y^3$ and $Y^4$ indicate NACA2412, NACA64$_1$-412, NACA65$_2$-415 and NACA64$_2$A215, respectively.

2. Results of direct optimization

To simplify the present aerodynamic optimization problem, low-speed airfoils are considered,

28

assuming that the flow field is governed by the two-dimensional, incompressible, inviscid flow equation. A simple panel method described in [28] can be used in the flow analysis.

Now let's consider the lift maximization problem. The objective function can be defined as the lift coefficient to be maximized. The only constraint used here is the maximum airfoil thickness set at 15% of the chord. The angle of attack is fixed at six degrees for this flow analysis.

First, lift maximization is considered with two design variables, $a_1$ for NACA2412 and $a_2$ for NACA64$_1$-412, for the demonstration purpose (thus $a_3 = a_4 = 0$). Since only two design variables are used, we can easily visualize the distribution of the objective function, lift coefficient $C_l$, as shown in Fig. 1 where $-5 < a_1, a_2 < 5$ and the lift coefficients are computed at intervals of 0.2 in both $a_1$ and $a_2$ coordinates. The negative value of the lift is replaced by zero.

The resulting distribution of the aerodynamic performance is highly irregular. Recall that both of the basis airfoils have 12% thickness. The design variables are expected to approximately satisfy the relation

$$a_1 + a_2 = \frac{15}{12} \tag{2}$$

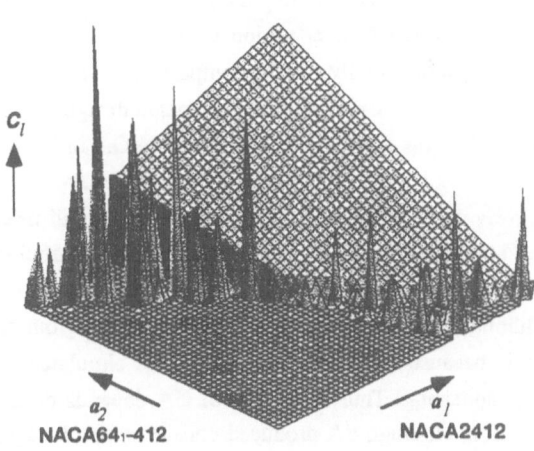

*Fig. 1 Lift distribution in design space*

29

to increase the airfoil thickness from 12% to 15%. However, the lift distribution is not smooth along the line of Eq. (2) because of the different cambers and thickness distributions of the two basis airfoils. In addition, the figure indicates that the maximum lift is achieved when $a_1 = a_2 = 5$. However, this is not acceptable because the airfoil thickness will be about 120% of the chord. Since the flow equation is linearized and only two design variables are involved, one might expect a smooth distribution of the objective function. On the contrary, the resulting distribution has sharp, distinct, multiple peaks. This is a typical situation where GM will not work.

To see the effect of the constraint, the objective function is now redefined by using a penalty function with the airfoil thickness to chord $t/c$ as

$$F = C_l \cdot \exp\left[-100 \times |t/c - 0.15|\right].$$  (3)

The corresponding function distribution is shown in Fig. 2. Although the distinct, multiple peaks are greatly reduced, most of the design space now has zero objective function value. The design space with positive objective function values is found only in the narrow ridge along with Eq. (2) and it still contains five local extrema. This is another typical situation where GM will have difficulty. In these situations, a mechanism to locate a global optimum is required. The combination of mutation and recombination of GA will be effective to find an optimal solution. On the other hand, the hill-climbing strategy of GM is practically inapplicable to this problem.

Now the next test case considers the lift maximization with a full set of four design variables. The same constraint and flow condition were used. This time three optimization methods, GM, SA and GA, were actually run for comparison. For GM, the feasible direction method in ADS V3.0 - a FORTRAN program for automated design synthesis [29] was used. For SA, the code listed in [2] was used. For GA, the simple GA in [4] was adapted with the real number coding.

Figure 3 summarizes the optimum lift values obtained from all three methods. For the GM and SA cases, four apparent initial designs are used. The lift coefficient of the GA case is 2.46, which is the best of the three and much higher than those of the previous two optimization cases. Although both GA and SA are capable of getting out of local extrema, GA outperforms SA. This is because GA uses a population for simulated evolution, while SA uses a single design for annealing. Thus, the result of GA depends on the initial design less than that of SA. In addition, although SA produced consistent results against different initial points, the results of GM strongly depend on the initial points. GM failed to find an optimum starting from the third initial design by reaching a negative thickness of the airfoil.

Figure 4 shows comparison of number of function evaluations required for the three methods. For the GM and SA cases, sum of four runs (corresponding to four different initial

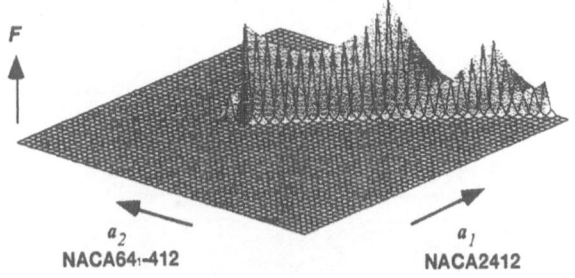

F

$a_2$
NACA64$_1$-412

$a_1$
NACA2412

*Fig. 2 Objective function distribution*
*(lift coefficient with a penalty for airfoil thickness)*

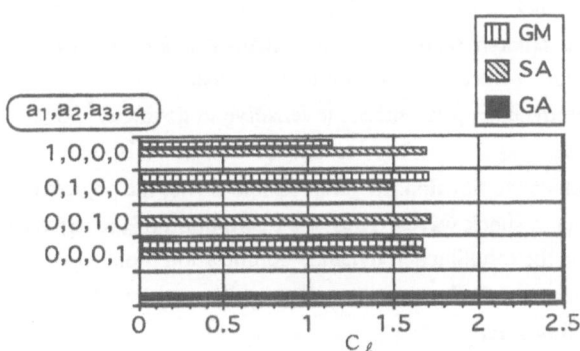

| | GM |
| | SA |
| ■ | GA |

$a_1, a_2, a_3, a_4$

1,0,0,0
0,1,0,0
0,0,1,0
0,0,0,1

0    0.5    1    1.5    2    2.5
$C_\ell$

*Fig. 3 Comparison of design results among GM, SA and GA*

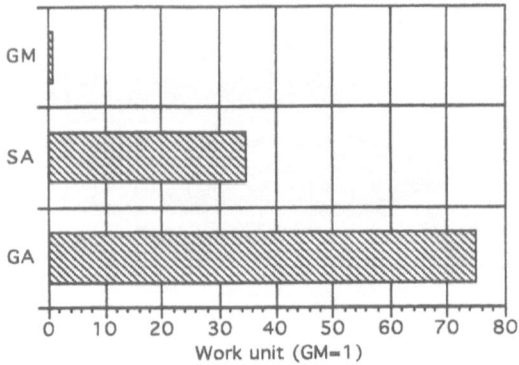

*Fig. 4 Comparison of number of function evaluations among GM, SA and GA*

designs) was plotted. It confirms that GA is the most time-consuming. However, if we use GM or SA, we have to run it starting from many, different initial designs to obtain a comparable result to that of GA. Since there are no guidelines on how to choose the initial design, the resultant random search must be exhaustive. Hence, GM and SA are poor in efficiency. Overall, GA is the best choice for this test case.

In general, aerodynamic performance is sensitive to geometry. So far, there is no theory in fluid dynamics that can tell us how to choose design variables that will guarantee the convexness of the objective function. In compressible flows, the objective function itself may be discontinuous due to shock waves. Thus, the distribution of the objective function will be quite unexpected and the resulting aerodynamic optimization problem will be very difficult. In this situation, a global search algorithm is indispensable and thus GA is the preferred choice for aerodynamic optimization.

## Genetic Algorithms

GAs are search algorithms based on the mechanics of natural selection and natural genetics. GAs work from a rich database of points (a population of strings), simultaneously climbing many peaks in parallel. Thus, the probability of finding a false peak is reduced with GAs as compared with the conventional methods that go from point to point like the gradient-based methods. In this section, various techniques used in GAs will be reviewed briefly.

1. Coding

In GAs, the natural parameter set of the optimization problem is coded as a finite-length string. Traditionally, GAs use binary numbers to represent such strings: a string has a finite length and each bit of a string can be either 0 or 1. For real function optimization, it is more natural to use real numbers. The length of the real-number string corresponds to the number of design variables.

As a simple test case, let's consider the following optimization:

Maximize: $\qquad\qquad\qquad\qquad f(x, y) = x + y$

Subject to: $\qquad\qquad\qquad x^2 + y^2 \le 1 \text{ and } 0 \le x, y \le 1$.

Let's represent the parameter set by using the polar coordinates here as

$$(x, y) = (r\cos\theta, \, r\sin\theta)$$

since the representation of the constraints will be simplified. Each point $(x, y)$ in the GA population is encoded by a string $(r, \theta)$. In the following test case, a population of 100 strings will be used where the initial strings are generated randomly.

## 2. SOGA

At each generation (iteration) of GA's process, fitness value (objective function value) of every individual is evaluated and used to specify its probability of reproduction. A new population is generated from selected parents by performing specific operators on their genes. These operators are briefly explained here.

Simple GA is composed of three operators: 1. Reproduction, 2. Crossover, 3. Mutation [4]. Reproduction is a process in which individual strings are copied according to their fitness values. This implies that strings with a higher fitness value have a higher probability of contributing one or more offsprings in the next generation. A typical reproduction operator is the roulette-wheel method described in Ref. 4. The reproduction process produces a mating pool. Then crossover proceeds in two steps. First, members in the mating pool are mated at random. Second, each pair of strings undergoes partial exchange of their strings at a random crossing site. This results in a pair of strings of a new generation. Mutation is a bit change of a string that occurs during the crossover process at a given mutation rate. Mutation implies a random walk through the string space and plays a secondary role in simple GA. A flowchart of the present GA is illustrated in Fig. 5.

A simple crossover operator for real number strings is the average crossover [30] which computes the arithmetic average of two real numbers provided by the mated pair. In this paper, a weighted average is used as

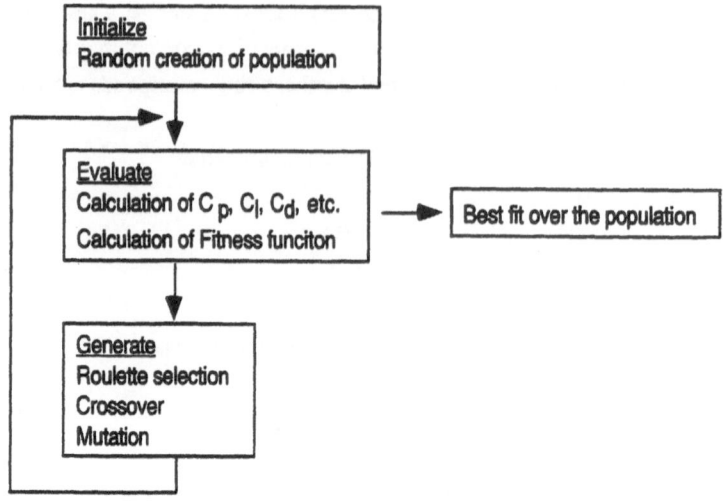

*Fig 5. Flowchart of GA*

$$Child1 = ran1*Parent1 + (1-ran1)*Parent2$$
$$Child2 = (1-ran1)*Parent1 + ran1*Parent2 \qquad (4)$$

where Child1,2 and Parent1,2 denote encoded design variables of the children (members of the new population) and parents (a mated pair of the old generation), respectively. The uniform random number $ran1$ in [0,1] is regenerated for every design variable. Because of Eq. (4), the number of the initial population is assumed even.

Mutation takes place at a probability of 20% (when a random number satisfies $ran2 < 0.1$). Equations (4) will then be replaced by

$$Child1 = ran1*Parent1 + (1-ran1)*Parent2 + m*(ran3-0.5)$$
$$Child2 = (1-ran1)*Parent1 + ran1*Parent2 + m*(ran3-0.5) \qquad (5)$$

where $ran2$ and $ran3$ are also uniform random numbers in [0,1] and $m$ determines the range of possible mutation. In the following test cases, $m$ was set to 0.4 for the radial coordinate $r$ and $\pi/3$ for the angular coordinate $\theta$.

## 3. Ranking

For a successful evolution, it is necessary to keep appropriate levels of selection pressure throughout a simulation [4]. Starting from a randomly created population, there is a tendency for a few superindividuals to dominate early on in the selection process. In this case, objective function values must be scaled back to prevent takeover of the population by these superstrings. When the population is largely converged, competition among population members is less strong and the simulation tends to wander. In this case objective function values must be scaled up to accentuate differences between population members to continue rewarding the best performers.

Scaling of objective function values has been used widely in practice [4]. However, this leaves the scaling procedures to be determined. To avoid such parametric procedures, a ranking method is often used [4]. In this method, the population is sorted according to objective function value. Individuals are then assigned an offspring count that is solely a function of their rank. The best individual receives rank 1, the second best receives 2, and so on. The fitness values are reassigned according to rank, for example, as an inverse of their rank values. Then the usual roulette-wheel method takes over with the reassigned values. The method described so far will be hereon referred to as SOGA (Single-Objective Genetic Algorithm).

## 4. MOGA

SOGA assumes that the optimization problem has (or can be reduced to) a single criterion (or objective). Most engineering problems, however, require the simultaneous optimization of multiple, often competing criteria. Such problems are called multiobjective (MO) or multicriteria problems. These problems have long attracted the attention of researchers using traditional techniques of optimization. More recently, GAs have been applied to the search for multicriteria optima [23].

The solution to multiobjective problems is often computed by combining multiple criteria into a single criterion according to some utility function. In many cases, however, the utility function is not well known prior to the optimization process. The whole problem should then be treated with non-commensurable objectives. As a second test case, let's consider the following optimization:

Maximize: $\qquad\qquad f_1 = x, \ f_2 = y$

Subject to: $\qquad\qquad x^2 + y^2 \leq 1 \text{ and } 0 \leq x, \ y \leq 1.$

Multiobjective optimization seeks to optimize the components of a vector-valued objective function. Unlike single objective optimization, the solution to this problem is not a single point, but a family of points known as the Pareto-optimal set. Each point in this set is optimal in the sense that no improvement can be achieved in one objective component that does not lead to degradation in at least one of the remaining components. The Pareto front of the present test case becomes a quarter arc of the circle $x^2 + y^2 = 1$ at $0 \leq x, \ y \leq 1$.

By maintaining a population of solutions, GAs can search for many Pareto-optimal solutions in parallel. This characteristic makes GAs very attractive for solving MO problems.

As solvers for MO problems, the following two features are desired: 1) the solutions obtained are Pareto-optimal and 2) they are uniformly sampled from the Pareto-optimal set. To achieve these with GAs, the following two techniques are successfully combined into MOGAs [25].

5. Pareto ranking

To search Pareto-optimal solutions by using MOGA, the ranking selection method described above for SOGA can be extended to identify the near-Pareto-optimal set within the population of GA. To do this, the following definitions are used: suppose $x_i = (x_i, y_i)$ and $x_j = (x_j, y_j)$ are in the current population and $f = (f_1, f_2)$ is the set of objective functions to be maximized,

1. $x_i$ is said to be dominated by (or inferior to) $x_j$, if $f(x_i)$ is partially less than $f(x_j)$, i.e., $f_1(x_i) \leq f_1(x_j) \wedge f_2(x_i) \leq f_2(x_j)$ and $f(x_i) \neq f(x_j)$.
2. $x_i$ is said to be non-dominated if there doesn't exist any $x_j$ in the population that dominates $x_i$.

Non-dominated solutions within the feasible region in the objective function space give the Pareto-optimal set.

Consider an individual $x_i$ at generation $t$ which is dominated by $p_i{}^t$ individuals in the current population (Fig. 6). Its current position in the individuals' rank can be given by

$$\text{rank}(x_i, t) = 1 + p_i'. \tag{6}$$

All non-dominated individuals are assigned rank 1 as shown in Fig. 6. The fitness assignment according to rank can be done similar to that in SOGA.

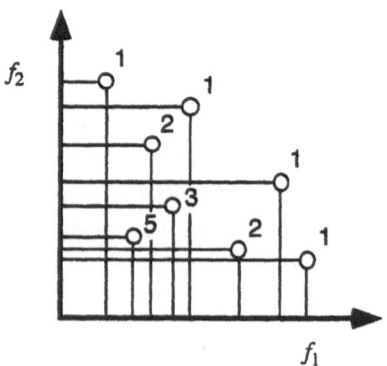

Fig. 6 Pareto ranking

36

## 6. Fitness sharing

To sample Pareto-optimal solutions from the Pareto-optimal set uniformly, it is important to maintain genetic diversity. It is known that the genetic diversity of the population can be lost due to their stochastic selection process. This phenomenon is called the random genetic drift [4].

Suppose that SOGA is applied to the objective function with equal peaks as explained in Ref. 4. In an initial population chosen uniformly at random, a relatively even spread of points across the function domain is obtained. As generations pass, the population climbs the hills, ultimately distributing most of the strings near the top of one hill among the others. The population converges to one peak or another without differential advantage. This is caused by the genetic drift, stochastic errors in sampling caused by small population sizes. To reduce the effect of these errors and to form stable subpopulations around each peak, the inducement of niche and species is considered.

To form the niche, the strings on a particular peak are forced to share the fitness value. If five strings are on a peak, each string receives only one fifth of the fitness value corresponding to the particular peak. The incorporation of forced sharing causes the formation of stable subpopulations (species) on different peaks (niches) in the problem.

A practical scheme is given by taking the raw fitness and dividing through by the accumulated number of shares

$$f_s(x_i) = \frac{f(x_i)}{\sum_j s(d(x_i, x_j))} \tag{7}$$

where $s(d)$ is a sharing function that determines the neighborhood and degree of sharing. The distance $d = d(x_i, x_j)$ can be measured with respect to a metric in either genotypic or phenotypic space. A genotypic sharing measures the interchromosomal Hamming distance. A phenotypic sharing, on the other hand, measures the distance between the designs' objective function values. In MOGAs, a phenotypic sharing is usually preferred since we seek a global tradeoff surface in the objective function space.

A typical sharing function is given by

$$s(d) = \begin{cases} 1 - \left(\frac{d}{\sigma_{share}}\right)^{\alpha} & d < \sigma_{share} \\ 0 & d \geq \sigma_{share} \end{cases} \tag{8}$$

This scheme introduces new GA parameters, the niche size $\sigma_{share}$ and the power $\alpha$. In the following, $\alpha$ is set to 0.1 to emphasize the niche count. Reference 25 gives a simple estimation of $\sigma_{share}$ in the objective function space. For the following test cases, it will be 0.02.

Niche counts can be consistently incorporated into the fitness assignment according to rank by using them to scale individual fitness within each rank. By implementing fitness sharing in MOGAs, one can expect to evolve a uniformly distributed representation of the global tradeoff surface.

## 7. Comparison of SOGA and MOGA

From the techniques described above, four optimization results are shown here for demonstration purposes. Figure 7 shows the result obtained from SOGA with real number coding. The GA population is represented by dots and the Pareto front of the MO test case is indicated by a solid arc. Since the solution of the present single-objective problem is $(1/\sqrt{2}, 1/\sqrt{2})$, many dots crowd near that point.

Figure 8 shows the result of the Pareto ranking. The Pareto front is recognized better by the population although the points still cluster in the middle of the front surface. Figure 9 shows the result of MOGA that combines the Pareto ranking and the fitness sharing. The population is distributed more uniformly along the Pareto front. On the other hand, the Pareto front receded slightly. This result suggests the use of elitism to preserve the Pareto solutions.

In Fig. 10, the best-$N$ selection [31] is incorporated further, where the best $N$ individuals are selected for the next generation among $N$ parents and $N$ children. The Pareto solutions will be kept once they are formed. As a result, the population grew to the Pareto set successfully as shown in the figure.

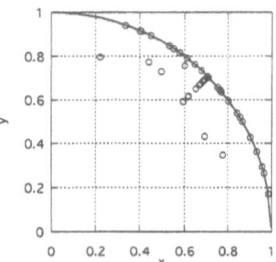

*Fig. 7 Result of SOGA*

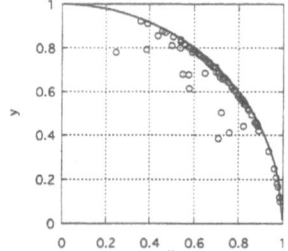

*Fig. 8 Result of MOGA*
*(Pareto ranking)*

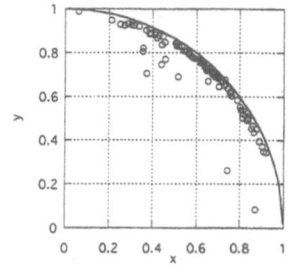 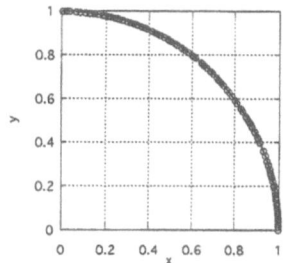

Fig. 9 Result of MOGA                    Fig. 10 Result of MOGA
(Pareto ranking and fitness sharing)        (plus best-N selection)

## Inverse optimization of airfoil

Target pressure optimization for airfoil is considered in this section. As shown in the previous section, GA will lead to the best solution for aerodynamic optimization. However, it requires tremendous computations. For engineering purposes, the direct approach will be too expensive. Therefore, the inverse optimization is developed to alleviate the computational time.

### 1. Coding

Genetic Algorithms simulate evolution by selection. Design candidates are considered as individuals in the population. An individual is characterized by genes represented as a string of parameters. Here an individual is a pressure distribution. Therefore, a coding scheme is required to specify a pressure distribution in terms of a string of parameters.

One of the parameterization techniques recommended for *ab initio* designs is B-spline parameterization [32]. The B-spline curve can be constructed such that the first and last points coincide with those of the defining polygon. Thus, pressure distributions are split into two curves, corresponding to pressure distributions on the upper and lower surfaces of an airfoil, respectively. As shown in Fig. 11, seven points are used to define a B-spline polygon, specifying pressure coefficient at the stagnation point $C_p = 1$ at $x = 0$ (since the inverse method does not solve the stagnation point exactly) and pressure coefficient at the trailing edge $C_p = C_{p,te}$ (by user specification) at $x = 1$ (assuming the chord length to be unity).

Among 14 points to define upper and lower $C_p$ curves, ten points are free to move. The $x$ and $y$ ($y = C_p$) coordinates of those points are the design variables and thus we have 20 design variables in total. Although standard GAs are characterized by the use of the binary

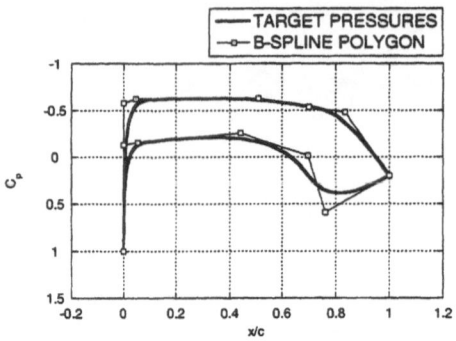

*Fig. 11 B-spline polygons and corresponding pressure distributions*

coding for design variables, a real (decimal) number is used for simplicity.

Initial population is generated randomly in the region of $0 < x < 1$ and $-1.5 < y < 1$. Besides the leading and trailing edges, one more point of the seven-point B-spline polygon is initially confined on the $y$ axis to describe a steep $C_p$ drop near the leading-edge region, that is, to obtain a large leading-edge radius typical for supercritical airfoil [33]. To create each individual, the lower $C_p$ curve is first generated based on constraints mentioned later. Then, the upper $C_p$ curve is generated so as to satisfy the rest of constraints approximately. If the resulting lift differs from the specified lift less than 10%, the corresponding string of parameters is assigned to an individual as genes. If it differs more than 10%, the string of parameters is discarded and regenerated. This process is repeated until 100 individuals are created.

## 2. SOGA

For the airfoil design, reduction of drag at a given lift is the primary objective. Thus, SOGA is adapted to optimization of pressure distribution around an airfoil. The two-dimensional airfoil drag can be broken down to skin friction, pressure drag due to separation, wave drag due to shock wave. To avoid separation in the inverse design, the maximum gradient of pressure coefficient should be restricted by empirical data. To avoid a shock wave, the target pressures should produce the shock-free distribution around the airfoil, too. They are enforced through the constraints of the present optimization. The remaining objective is the minimization of the skin-friction drag.

In this approach, the fitness scaling was used instead of the ranking method so that the maximum fitness became 1 and the minimum fitness became 0 at every generation. In addition, the universal stochastic sampling method was used instead of the simple roulette wheel selection.

## 3. Fitness evaluation: objective and constraints

In this paper, we define the optimization problem as

Minimize:       Drag coefficient $C_d$

Subject to:     1. Lift coefficient $C_l$ = specified

2. Airfoil thickness $t$ = specified

3. $C_{p,l} < 0$ at $0.1 < x < 0.6$ and $\int_{0.1}^{0.6} |C_{p,l}| dx \geq 0.1$

4. $\max_{0.6 < x < 1} C_{p,l} < 0.4$

5. $C_p \big|_{\text{suction peak}} \leq C_p^*$

6. $\dfrac{dC_{p,u}}{dx} \bigg|_{0.1 < x < 0.5} \approx 0$

7. $\dfrac{dC_{p,u}}{dx} \leq 2.5$

8. Number of inflection points < 2 .

To specify airfoil thickness $t$ approximately, a formula is taken from [18] as

$$t = -\frac{\sqrt{1 - M_\infty^2}}{2} \int_0^1 \frac{C_{p,u} + C_{p,l}}{2} dx \qquad (9)$$

where $M_\infty$ is the freestream Mach number. The constraints 5 and 6 aim to reproduce sonic-plateau pressure distributions described in [33]. The constraint 7 was taken from the observation of $C_p$ plots in [33] to avoid a separation of the boundary layer.

Drag is calculated as a sum of viscous and wave drag. When the flow is attached, the profile drag can be calculated from knowing the potential-flow pressure distributions and location of transition from laminar to turbulent flow. Locations of transition are left for user specification. When no shock wave is present, wave drag is ignored.

The Squire-Young relation is an empirical relation between drag based on the momentum thickness of the boundary layer and the potential velocity [34]. The momentum thickness can be estimated from an integral equation of the turbulent boundary layer. The viscous drag is estimated as

$$C_{d_v} = 2\theta_{te} \left( \frac{U_{e,te}}{U_\infty} \right)^{3.2} \qquad (10)$$

where $\theta_{te}$ is the momentum thickness at the trailing edge, $U_{e,te}$ is the potential velocity at the trailing edge and $U_\infty$ is the freestream velocity. See [34] for more detail.

The optimization took about three minutes on an SGI Indy workstation using 1000 successive generations with 100 individuals in the population. It is far less than the computational time necessary for the inverse design as will be mentioned later.

## 4. Construction of fitness function

GAs require an objective to be maximized. An inverse of the drag coefficient is taken as the objective here. Then, all the constraints are required to be combined with the objective. There are eight constraints in Fitness Evaluation section. The constraints 1 and 2 are equality constraints and the constraints 3 to 8 are inequality constraints. The equality constraints are multiplied to the objective as an exponential function to reject the individuals that do not satisfy the specified lift and airfoil thickness. The inequality constraints are expressed so as to increase their values when violated and to add the inverse of their sum to the objective with penalty.

The final form of the objective is given as

$$Fitness = \left( \frac{0.4}{C_d^2} + \frac{2 \times 10^7}{IC^2} \right) \cdot \exp(-100 \times EC) \tag{11}$$

where $IC$ and $EC$ denote the sum of inequality constraints and equality constraints, respectively. For $IC$, the constraints 3 to 8 are represented as

$$
\begin{aligned}
IC = 10000 &\times \left[ \min\left( \int_{0.1}^{0.6} |C_p| dx,\ 0.1 \right) - 0.1 \right]^2 \\
+5 &\times \left[ \max_{0.6 < x < 1} \left( C_{p,l},\ 0.4 \right) - 0.4 \right]^2 \\
+3 &\times \left[ \max\left( \frac{C_p|_{\text{suction peak}}}{C_p^*},\ 1.0 \right) - 1.0 \right]^2 \\
+40000 &\times \left[ 1.1 - \min\left( \frac{slope + |slope|}{2},\ 1.1 \right) \right] \\
+0.001 &\times \left[ \max\left( \frac{dC_{p,u}}{dx},\ 2.5 \right) - 2.5 \right]^2 \\
+0.01 &\times \exp[\text{Number of inflection points}]
\end{aligned}
\tag{12}
$$

where $slope = \dfrac{C_{p,u}|_{x=0.5} - C_{p,u}|_{x=0.1}}{0.4} + 1$. For $EC$, the constraints 1 and 2 are represented as

$$EC = \max\left(\left|t_{\text{specified}} - t_{\text{calculated}}\right|, 10^{-4}\right)$$
$$+ \max\left(\left|C_{l\,\text{specified}} - C_{l\,\text{calculated}}\right|, 10^{-4}\right) \qquad (13)$$
$$-0.01$$

where the differences below 10⁻⁴ are ignored for the optimization.

## 5. Inverse design cycle

Once the present GA finds an optimum target pressure distribution, a corresponding airfoil geometry can be obtained by an inverse design method (Fig. 12). Here the inverse design code WinDes, is used [19]. The code can solve both two- and three-dimensional problems.

WinDes uses the following iterative procedure. Suppose the initial geometry and surface pressure distributions obtained from any CFD analysis code are given. First, pressure differences are calculated from the given initial and target pressure distributions. From these pressure differences, corresponding geometry corrections can be computed from the integral equations discretized at the panels on the geometry. An improved geometry is then obtained from the initial geometry and the computed geometry corrections. Finally, the CFD code is used again to check how close the resulting pressure distributions are to the target distributions. If the differences are still large, the process will be iterated. In practice, 10 to 20 iterations are sufficient to obtain the final geometry.

The advantage of this method is that the analysis code required is arbitrary and any type of analysis method, even experimental, can be used. In this paper, two- and three-dimensional Navier-Stokes codes, LANS2D [35] and LANS3D [20], were used. The latest version of these codes uses the third-order upwind in the right-hand side.

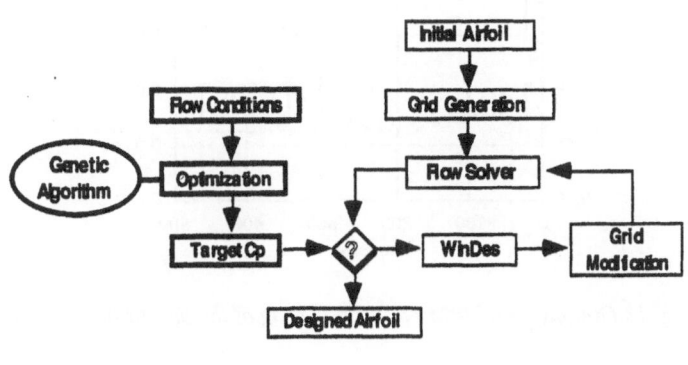

Fig. 12 Flowchart of the present design procedure

43

In the present inverse design, grid generation around the modified geometry is required at every iteration. In order to automate the inverse design loop, the grid generator must be made robust and efficient. An algebraic grid generation code described in [35] is used. The two-dimensional C-type mesh contains 131 times 51 grid points in the chordwise and normal (to the surface) directions, respectively. In the three-dimensional case, the C-H topology is used, applying the two-dimensional grid generation at each spanwise section. The three-dimensional grid contains 30 sections in the spanwise direction. Use of WinDes, LANS2D /3D, and the algebraic grid generator constructs an automated loop for the inverse design with reasonable computational requirements.

## 6. Results of airfoil design

In this optimization, flow condition was set to the freestream Mach number of 0.75 and the Reynolds number of 10 million. In the Navier-Stokes computation, the Baldwin-Lomax turbulence model [36] was used. The angle of attack was set to zero. Points of transition were fixed at 5 and 10% chord for upper and lower surfaces, respectively. The lift was specified as 0.5 and the trailing-edge pressure coefficient was set to 0.15. Figure 13 shows the optimization history of the present GA. The optimum was obtained after about 450 generations.

Figure 14 shows the target pressure distribution obtained from the present GA, the

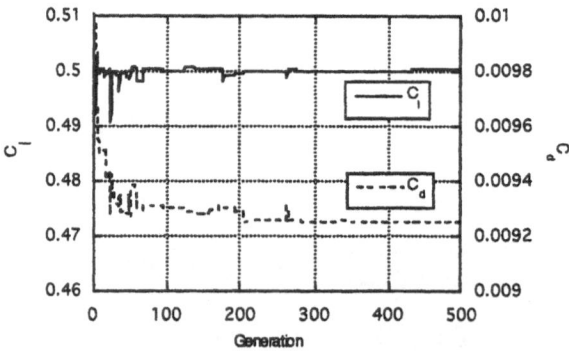

*Fig. 13 Optimization history of lift and drag of the best fit in generation*

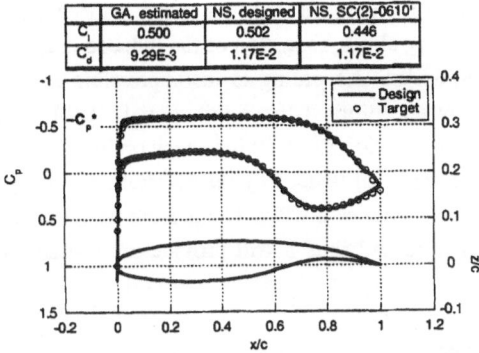

| | GA, estimated | NS, designed | NS, SC(2)-0610' |
|---|---|---|---|
| $C_l$ | 0.500 | 0.502 | 0.446 |
| $C_d$ | 9.29E-3 | 1.17E-2 | 1.17E-2 |

*Fig. 14 Design result and comparison of aerodynamic performance of airfoil case*

designed geometry obtained from the inverse method, and the corresponding pressure distribution obtained from the Navier-Stokes computation. The optimized pressure distribution has a sonic plateau to avoid a shock wave on the upper surface of the airfoil and a rear loading region typical for the supercritical airfoils. In the inverse design cycle, initial geometry was chosen as NACA0012 airfoil. In total, nine iterations were required to obtain the final geometry.

In Fig. 14, the attached table shows a comparison of the aerodynamic performances among the target (indicated as GA, estimated), the inverse design result (indicated as NS, designed) and a modified supercritical airfoil (indicated as NS, SC(2)-0610'). The SC(2)-0610 airfoil is taken from [33]. Although it has a blunt trailing edge, the geometry is modified to have a sharp trailing edge for a comparison purpose. The drag obtained from the target optimization was underestimated due to the simplified calculation of the viscous drag. Comparing the NS (Navier-Stokes) results, the present design shows higher $L/D$ (lift-to-drag) ratio than the modified SC(2)-0610 airfoil.

## Inverse Optimization of Wing

Once the airfoil shape is designed, the next step of the wing design process is to determine the variation of the designed airfoil in the spanwise direction. The design principles for this step are essentially twofold as mentioned in the beginning of this chapter. One is to preserve the two-dimensional performance as much as possible. This is easily achieved by the inverse method by specifying the same chordwise pressure distribution along the wing span. The resulting wing has the straight isobar pattern of pressure contours on the wing surface.

The other is to minimize the induced drag. Since the induced drag becomes one half to two third of the total drag during climb, reduction of the induced drag is an important goal for the three-dimensional wing design. According to incompressible flow theory, the minimum induced drag is achieved by an elliptical lift distribution [22]. Therefore, the elliptical lift distribution is the ideal design criterion for the wing shape optimization.

The induced drag is strongly influenced by the planform of a wing. Planform is directly related to aspect ratio and taper ratio of a wing. The present design method will minimize the induced drag for any taper ratio and thus it will provide more design opportunity for wing shapes.

1. Pressure distribution for wing

Target pressure distribution for the three-dimensional wing can be obtained by specifying the chordwise pressure distributions at several spanwise sections. Planform shape of a wing is usually determined by other means and thus a typical wing planform of transonic transport aircraft is assumed here.

The present objective of the wing design is to minimize the induced drag. This is achieved by the elliptical lift distribution in the spanwise direction of the wing. The constraint in the total lift will specify an elliptical lift distribution uniquely. Thus, the objective function can be given by differences of the sectional lifts to the elliptic distribution at the several spanwise sections. The three-dimensional optimization problem is now defined as

Minimize: 1. Difference of the spanwise lift distribution to the elliptic distribution
2. Two-dimensional drag coefficient $C_d$ at each spanwise section
Subject to: 1. Additional constraints for chordwise pressure distribution at each spanwise section
2. Straight isobar pattern of pressure contours .

We can further redefine the constrained problem to the multiobjective optimization problem as

Minimize: 1. Difference of the spanwise lift distribution to the elliptic distribution
2. Two-dimensional drag coefficient $C_d$ at each spanwise section
3. Penalty function for chordwise pressure distribution at each spanwise section as described in the previous section.
Subject to: 1. Straight isobar pattern of pressure contours .

2. MOGA

Before implementing the Pareto ranking approach for the present MOGA, we have tried a few other ways to construct a GA for the present multiobjective optimization. First, a SOGA was used by combining three objective functions into a single one. However, this approach not only failed to search Pareto-optimal solutions, but also produced premature convergence. Certain spanwise airfoil sections had unacceptable chordwise pressure distributions. Next, the Vector Evaluated Genetic Algorithm (VEGA) [24] was adapted to the present problem. As pointed out in [23], however, the solution was extremely good for one objective but not for the others. These experiences led us to Fonseca-Fleming's Pareto ranking method [25].

In the present MOGA, the third objective for the penalty function is used to pool the top 30% individuals in the population. Then Fonseca-Fleming's Pareto ranking method is applied to these individuals by using the first and second objectives. Selection operator is defined by using the nonlinear function suggested in [37]. The elite strategy [4] is also used to preserve the best individual for each objective instead of the best-N selection. After 200 generations, the best solution in terms of the first objective is selected from the Pareto-optimal set as the optimal solution.

As mentioned in the previous section, random creation of initial population produces infeasible solutions due to the severe constraints. Thus, we first ran the two-dimensional GA by using only the constraints to evolve a population of feasible solutions. Then we distributed the sectional pressure distribution to the six spanwise sections from the root to the 83.3% span so as to approximate the elliptical lift distribution. To do this, we only changed the pressure on the lower surface of the airfoil. In this way, we were able to implicitly satisfy the first design criterion for the wing mentioned in the beginning, that is, to maintain the two-dimensional performance. The straight isobar pattern of pressures on the upper surface of the wing is expected to produce the drag divergence at the same Mach number along the wing span and thus the resulting drag-divergence Mach number of the wing will be similar to that of the airfoil section. As the initial population of the present MOGA, 210 individuals were used.

Once the present MOGA finds an optimum target pressure distribution, corresponding wing geometry can be obtained by the inverse design method similar to the airfoil case. The inverse design code, the Navier-Stokes code, and the algebraic grid generator were implemented on a NEC SX-4 supercomputer at Department of Aeronautics and Space Engineering, Tohoku University. The inverse design for one cycle required about 45 min of single CPU time (mostly used for Navier-Stokes computation).

3. Results of wing design

As a model wing for transonic transport aircraft, a simple, swept and tapered wing shown in the left-hand side of Fig. 15 is considered for shape optimization. The wing has a sweep angle of 20.4 deg, an aspect ratio of 7.77 and a taper ratio of 0.3. There is a trailing-edge kink at the 33.3% semispan section similar to practical designs.

The elliptical lift distribution is monitored at six locations from the root to the 83.3% span as indicated. The inverse solver uses the same spanwise locations for the geometry correction. For the Navier-Stokes grid, the modification of wing geometry was linearly interpolated between those sections. In the tip region, the same airfoil section was used outside of the 83.3% section, while the wing twist was linearly extrapolated. The tip region is usually designed separately by other means and thus the optimization of this region is not considered here.

The right-hand side of Fig. 15 shows the computed pressure contours on the upper surface of the wing designed by the inverse method based on the target pressure distribution optimized by the present MOGA. Flow condition was at freestream Mach number of 0.8, Reynolds number based on the root chord of $10^7$ and an angle of attack of 0 deg. The resulting straight isobar pattern satisfies the first design principle well and thus indicates good performance at higher Mach numbers. On the other hand, it shows minor oscillation near the leading edge toward the root section. Although the airfoil sections vary a lot from the root to 16.7 % section, a linear interpolation is used to create a Navier-Stokes grid for brevity. To treat the root region as well as the tip region more precisely, an elaborated procedure may be necessary. In addition, the specification of the straight isobar pattern up to the wing root contradicts the symmetry boundary condition at the root section used in the Navier-Stokes solver.

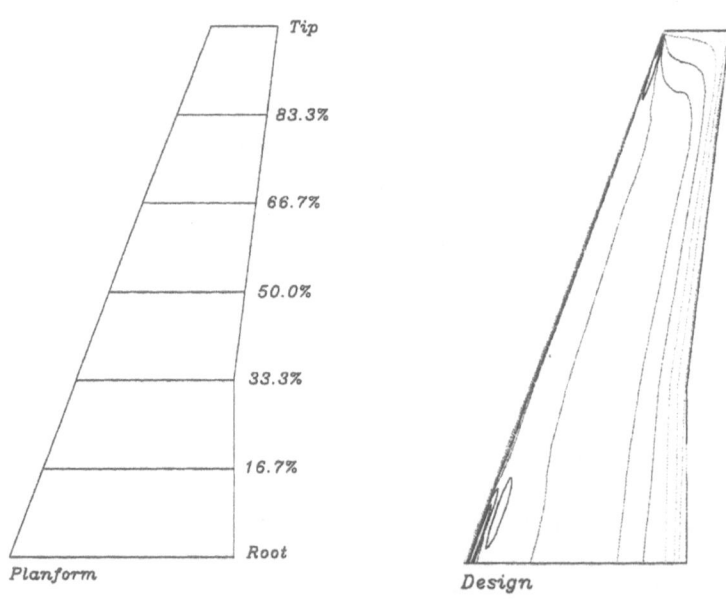

*Fig. 15 Wing planform and computed pressure distribution on the designed wing*

Figure 16 shows the computed lift distribution of the designed wing in comparison with the elliptic distribution. The result is found to be in close agreement with the second design principle closely. Figure 17 shows the target chordwise pressures obtained from the present MOGA, the resulting airfoil shape of the wing and the corresponding pressures computed by the Navier-Stokes solver at the 16.7%, 50.0% and 83.3% spanwise sections. It confirms that the inverse problem is solved satisfactorily except at the leading edge near the root section. The discrepancy of the pressure profiles there corresponds to the oscillation found in Fig. 15.

Figure 18 summarizes the aerodynamic performance of the designed wing in terms of the lift-to-drag ratio. The present design has $C_L = 0.489$ and $C_D = 0.0195$ at the design point of $M_\infty = 0.8$. The calculated drag divergence Mach number is $M_{DD} = 0.818$ at $C_D = 0.0208$.

For comparison purposes, another wing was designed by the inverse method. The design indicated as 'Isobar' in the figure was obtained by specifying the straight isobar pattern on both upper and lower surfaces of the wing. This is in fact a standard design procedure for transonic wings. The resulting wing satisfies the first design principle of the wing exactly but not the second one, and thus, it is expected to produce more induced drag than the present design obtained from MOGA. The isobar design has $C_L = 0.506$ and $C_D = 0.0207$ at the design point of $M_\infty = 0.8$. The calculated drag divergence Mach number is $M_{DD} = 0.817$ at $C_D = 0.0220$.

The $L/D$ plots confirm that the present design has a better performance over a wide range of Mach numbers than the isobar design because of the reduction of the induced drag through the elliptic loading. On the other hand, in the isobar design, reduction of the induced drag is dependent on finding the proper taper ratio. Since the present taper ratio is 0.3, the resulting wing is expected to give fairly good performance nevertheless.

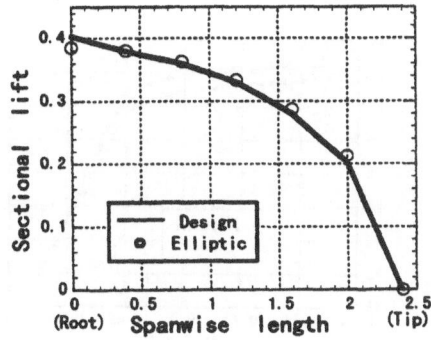

Fig. 16 Sectional lift distribution in the spanwise direction

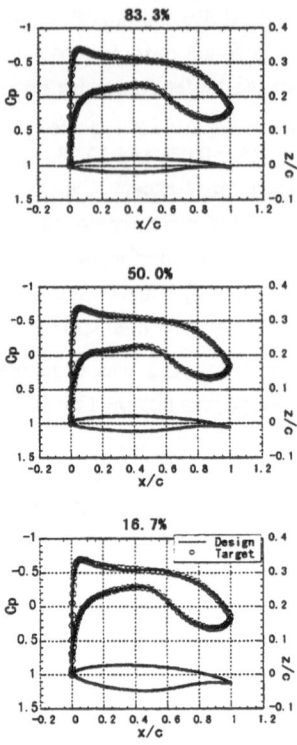

Fig. 17 Designed airfoil sections and corresponding pressure distributions

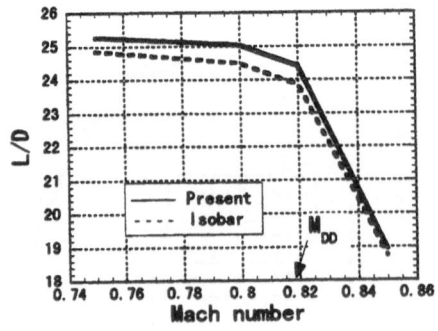

Fig. 18 Comparison of L/D performances

# Conclusion

Characteristics of aerodynamic optimization have been discussed through wing shape design problems. It is demonstrated that the distribution of the objective function can be extremely rough even in a simplified problem. In such situations, GA (Genetic Algorithm) is expected to be more effective than a simple hill-climbing strategy.

Three optimization algorithms, the GM (gradient-based method), SA (Simulated Annealing) and GA, were first applied to the airfoil shape design using the approximation concept to compare their performances. Although GA is time-consuming, its result is superior to those of the other two. Since the other algorithms require many trials starting from various initial designs to obtain a comparable result, they are less efficient. The results suggest that GA is the best choice for aerodynamic optimization.

To alleviate large computational time necessary for GA, the inverse optimization method has been developed to optimize target pressure distributions. GA is applied to find a pressure distribution that minimizes the airfoil drag under constraints on lift, airfoil thickness and other design principles. Once the target pressure is given, the corresponding geometry can be found by an inverse code coupled with a Navier-Stokes solver.

MOGA based on Fonseca-Fleming's Pareto ranking method has been developed to optimize the three-dimensional target pressures for the aerodynamic inverse design of wing shape. The optimization problem was formulated to minimize the induced drag for wing as well as to minimize the viscous drag for airfoil sections.

The resulting procedure was successfully applied to transonic wing design. The standard design procedure for transonic wings was previously focused on materializing the straight isobar pattern over the wing. Reduction of the induced drag was dependent merely on the use of a proper taper ratio for the wing planform. The present design procedure allows the minimization of the induced drag for arbitrary wing planform with any taper ratio. This will provide more design opportunities for wing shapes in terms of better aerodynamic performance, lighter structural wight, and less expensive manufacturing cost.

# References

[1]  van den Dam, R. F., van Egmond, J. A. and Slooff, J. W.: Optimization of target pressure distributions, Special Course on Inverse Methods for Airfoil Design for Aeronautical and Turbomachinery Applications, AGARD Report No. 780, Reference 3, (1990).

[2]  Press, W. H., *et al.*: *Numerical Recipes in FORTRAN: the art of scientific computing*, 2nd ed., Cambridge University Press, Cambridge, (1992).

[3]  Vanderplaats, G. N.: *Numerical Optimization Techniques for Engineering Design: with applications*, McGraw-Hill, Inc., New York, (1984).

[4]  Goldberg, D. E.: *Genetic Algorithms in Search, Optimization & Machine Learning,*

Addison-Wesley Publishing Company, Inc., Reading, (1989).

[5] Bramlette, M. F. and Cusic, R.: A comparative evaluation of search methods applied to the parametric design of aircraft, Proceedings of the Third International Conference on Genetic Algorithms, Morgan Kaufmann Publishers, Inc., San Mateo, pp. 213-218, (1989).

[6] Crispin, Y.: Aircraft conceptual optimization using simulated evolution, AIAA Paper 94-0092, (1994).

[7] Powell, D. J., Tong, S. S. and Skolnick, M. M.: EnGENEous domain independent, machine learning for design optimization, Proceedings of the Third International Conference on Genetic Algorithms, Morgan Kaufmann Publishers, Inc., San Mateo, pp. 151-159, (1989).

[8] Gage, P. and Kroo, I.: A role for genetic algorithms in a preliminary design environment, AIAA Paper 93-3933, (1993).

[9] Gregg, R. D. and Misegades, K. P.: Transonic wing optimization using evolution theory, AIAA Paper 87-0520, (1987).

[10] Quagliarella, D. and Cioppa, A. D.: Genetic algorithms applied to the aerodynamic design of transonic airfoils, AIAA Paper 94-1896, (1994).

[11] Yamamoto, K. And Inoue, O.: Applications of genetic algorithm to aerodynamic shape optimization, AIAA Paper 85-1650-CP, A collection of technical papers, 12th AIAA Computational Fluid Dynamics Conference, CP956, San Diego, CA, pp. 43-51, (1995).

[12] Obayashi S. and Takanashi, S.: Genetic optimization of target pressure distributions for inverse design methods, *AIAA Journal*, vol. 34, no. 5, pp. 881-886, (1996).

[13] Doorly, D.: Parallel genetic algorithms for optimization in CFD, *Genetic Algorithms in Engineering and Computer Science*, Winter, G., *et al.* (ed.), John Wiley & Sons, Chichester, pp. 251-270, (1995).

[14] Périaux, J., *et al.*: Robust genetic algorithms for optimization problems in aerodynamic design, *Genetic Algorithms in Engineering and Computer Science*, Winter, G., *et al.* (ed.), John Wiley & Sons, Chichester, pp. 371-396, (1995).

[15] Poloni, C.: Hybrid GA for multi objective aerodynamic shape optimization, *Genetic Algorithms in Engineering and Computer Science*, Winter, *et al.* (ed.), John Wiley & Sons, Chichester, pp. 397-416, (1995).

[16] Labrujére, Th. E. and Slooff, J. W.: Computational methods for the aerodynamic design of aircraft components, *Annual Review of Fluid Mechanics*, Vol. 25, pp.183-214, (1993).

[17] van den Dam, R. F.: Constrained spanload optimization for minimum drag of multi-lifting surface configurations, Computational Methods for Aerodynamic Design (Inverse) and Optimization, AGARD Conference Proceedings No. 463, Reference 16, (1990).

[18] van Egmond, J. A.: Numerical optimization of target pressure distributions for subsonic and transonic airfoil design, Computational Methods for Aerodynamic Design (Inverse) and Optimization, AGARD Conference Proceedings No. 463, Reference 17, (1990).

[19] Takanashi, S.: Iterative three-dimensional transonic wing design using integral equations,

*Journal of Aircraft*, Vol. 22, No. 8, pp. 655-660, (1985).

[20] Fujii, K. and Obayashi, S.: Navier-Stokes simulations of transonic flows over a practical wing configuration, *AIAA Journal*, Vol. 25, No. 3, pp. 369-370, (1987).

[21] Torenbeek, E.: *Synthesis of Subsonic Airplane Design*, Kluwer Academic Publishers, Dordrecht, pp. 215 -262, (1982).

[22] Anderson, Jr., J. D.: *Introduction to Flight*, McGraw-Hill Inc., New York, pp. 216-222, (1989).

[23] Tamaki, H., Kita H. and Kobayashi, S.: Multi-objective optimization by genetic algorithms: a review, Proceedings of 1996 IEEE International Conference on Evolutionary Computation, pp. 517-522, (1996).

[24] Schaffer, J. D.: Multiple objective optimization with vector evaluated genetic algorithm, Proceedings of the 1st International Conference on Genetic Algorithms, Morgan Kaufmann Publishers, Inc., San Mateo, pp. 93-100, (1985).

[25] Fonseca C. M., and Fleming, P. J.: Genetic algorithms for multiobjective optimization: formulation, discussion and generalization, Proceedings of the 5th International Conference on Genetic Algorithms, Morgan Kaufmann Publishers, Inc., San Mateo, pp. 416-423, (1993).

[26] Horn, J., Nafplitois, N. and Goldberg, D., E.: A niched Pareto genetic algorithm for multiobjective optimization, Proceedings of the 1st IEEE Conference on Evolutionary Computation, pp. 82-87, (1994).

[27] Michielessen, E. and Weile D. S.: Electromagnetic System Design Using Genetic Algorithms, *Genetic Algorithms in Engineering and Computer Science*, Winter, *et al.* (ed.), John Wiley & Sons, Chichester, pp. 345-370, (1995).

[28] Katz, J. and Plotkin, A.: *Low-Speed Aerodynamics: from wing theory to panel methods*, international edition, McGraw-Hill Inc., New York, (1991).

[29] Vanderplaats, G. N.: ADS – A FORTRAN program for automated design synthesis, Version 3.00, Engineering Design Optimization, Inc., (1988).

[30] Davis, L.: *Handbook of Genetic Algorithms*, Van Nostrand Reinhold, (1990).

[31] Tsutsui, S. and Fujimoto, Y.: Forking genetic algorithms with blocking and shrinking modes (fGA), Proceedings of the 5th International Conference on Genetic Algorithms, Morgan Kaufmann Publishers, Inc., San Mateo, pp. 206-213, (1993).

[32] Rogers, D. F. and Adams, J. A.: *Mathematical Elements for Computer Graphics*, Second Edition, McGraw-Hill, Inc., New York, (1990).

[33] Harris, C. D.: NASA supercritical airfoils – a matrix of family-related airfoils, NASA TP-2969, (1990).

[34] Young, A. D.: *Boundary Layers*, AIAA Education Series, Washington, D. C., (1989).

[35] Matsushima, K., Obayashi, S. and Fujii, K.: Navier-Stokes computations of transonic flow using the LU-ADI method," AIAA Paper 87-0421, (1987).

[36] Baldwin, B. S. and Lomax, H.: Thin-layer approximation and algebraic model for separated turbulent flows, AIAA Paper 78-257, (1978).

[37] Michalewicz, Z.: *Genetic Algorithms + Data Structures = Evolution Programs*, Second extended edition, Springer-Verlag, Berlin, pp. 57-58, (1994).

# Subsonic Aerodynamic Design Via Optimization

Krzysztof Kubrynski

Institute of Aeronautics and Applied Mechanics,
Warsaw University of Technology, ul.Nowowiejska 24, 00-665 Warsaw, POLAND
e-mail: kkubryn@meil.pw.edu.pl

### Summary

The paper presents a panel method which allows to design 3-dimensional configuration with prescribed pressure distribution at specified design angles of attack. Effects of flaps deflection can be incorporated and the induced drag can be minimized. A higher order panel method is applied to perform flow analysis. Flaps deflection is simulated by modifying a Neumann boundary conditions. In the design case the geometry which minimizes differences between the design and actual pressure distributions and/or minimizes induced drag is found iteratively using an optimization technique. Geometrical constrains and regularity conditions can be specified by means of penalty function concept and the requested values of lift and moment coefficients can be enforced using Lagrange multipliers technique.

KEY WORDS: aerodynamic design, maneuver flaps design, induced drag minimization

## Introduction

Computational methods play an important role in aerodynamic development of modern flying vehicles. This is especially true in case of high speeds (mainly transonic). Low speed aerodynamics more and more frequently uses advanced methods - which significantly improves performance. In order to improve the aerodynamic efficiency many assistant devices (e.g. leading/trailing edge flaps or winglets) are applied. There exist many methods of flow analysis used to analyze characteristics of such aerodynamic configurations, but inverse and optimization methods can significantly improve the process of aerodynamic design and final aerodynamic efficiency. During the process of aerodynamic design of the aircraft, the external shape is defined in such a way that it satisfies performance requirements at a single design point (e.g. cruise conditions of airliner) or at a number of design points (if the plane is intend to be equally good at low and high speeds, at high $n$ maneuvers or high altitudes etc.). Additional characteristics at off-design conditions (like stall or dive) must be also considered. In the case of a wide range of performance requirements, there may exist some incompatibility in the design. The high CL requirements (low speed, maneuver etc.) are always in conflict with those at low CL (e.g. sea level dash). This is especially true for the case of a thin wing - used due to its low friction drag and high critical Mach number at design angle of attack, and for high performance gliders as well. One way to overcome this conflict is to utilize variable geometry of the wing sections or the planform. The simplest case of such a solution is the use of the leading edge and/or trailing edge flaps which allows for approximation the

55

optimal geometry at different flight conditions.  In the case of low speed aerodynamics there are three basic problems which influence the final efficiency of the plane and must be carefully considered:  friction drag, induced drag and interference (and the corresponding interference drag). The problem of optimal design of an aircraft is not easy, especially in the case of 3-dimensional complex configuration, where interference effects occur between close coupled elements.  Assembly of suitably designed separate elements usually leads to adverse interference effects and loss of efficiency.  In principle it is possible to design a configuration with neutral, or even favorable interference, where the interaction between the aircraft components leads to better global characteristics then those of separate elements, reducing friction and pressure drag.  The most common method of designing such a configuration is the direct multi-point optimization with a set of objective functions (e.g. drag, L/D ratio etc.),  constrains on geometrical (e.g. wing thickness) and aerodynamic (e.g. CL, Cm0) origin and parameters defining geometry and flaps position as independent variables. This conceptually attractive method is not easy to apply in the case of 3-dimensional real configurations because of its cost and usually insufficient accuracy of the available methods of flow analysis. Finding real solutions is often impossible.  On the other hand, the flow characteristics (especially boundary layer development and friction drag) at given conditions (Reynolds and Mach numbers) depend on distribution of the surface pressure.  The above mentioned design problem can be solved in two steps.  In the first, a distribution of a surface pressure at various design points is specified in such a way that it satisfies aerodynamic requirements (e.g. extended region of laminar flow for low friction drag or no separation - found by inverse boundary layer method).  In the second step the geometry which produces distribution of surface pressure closest to the target one.is found using multi-point inverse method.  The paper presents a method which allows to solve the second problem in the case of  linearized subcritical flow (including weak viscous-inviscid interaction) and complex geometry. Additionally, the induced drag can be minimized directly at design angles of attack. The method is an extended and developed version of the previous design methods of the author [1,2].

## Flow analysis

The analysis method is based on the linearized theory [3] of compressible subsonic flow. The Prandtl-Glauert equation is assumed to govern the perturbation velocity potential in the flow field:

$$\beta^2 \varphi_{xx} + \varphi_{yy} + \varphi_{zz} = 0 \quad ; \quad \beta^2 = 1 - Ma_\infty^2 . \tag{1}$$

The linearized mass flux boundary conditions are applied on the external surface:

$$\mathbf{W} \cdot \mathbf{n} = (\mathbf{V}_\infty + \mathbf{w}) \cdot \mathbf{n} = \overset{\circ}{m}_s / \rho_\infty \tag{2}$$

where $\mathbf{w}$ is the linearized perturbation mass flux vector and $\overset{\circ}{m}_s$ is the intensity of mass outflow through the surface.  A second order pressure formula is applied to find aerodynamic forces and moments and an isentropic formula to express the surface distribution of pressure. Using Green's Theorem the perturbation velocity potential is expressed as:

$$E_P \varphi_P = \oiint_{S_B} \left[ \frac{-(\dot{m}_\infty / \rho_\infty - \mathbf{V}_\infty \cdot \mathbf{n}_Q)}{4\pi r_\beta} + \varphi_Q \beta^2 \frac{\mathbf{r} \cdot \mathbf{n}_Q}{4\pi r_\beta^3} \right] dS + \oiint_{S_W} <\varphi>_Q \beta^2 \frac{\mathbf{r} \cdot \mathbf{n}_Q}{4\pi r_\beta^3} dS \quad (3)$$

where $<\varphi>$ is the jump of the potential across the wake and E is a function of position (1, 1/2 and 0 for P in the flow field, on the surface and outside the flow field respectively). Equation (3) is solved by a panel method based on quadratic doublets, linear source distribution and indirect Dirichlet boundary conditions (zero perturbation potential is specified on internal side of the body surface). Control points and unknown singularity parameters (one per panel) are located at a panel's center of gravity. The jump of potential across the wake behind the lifting surface is determined by the Kutta condition: vanishing velocity component in the direction normal to trailing edge bisector or to wake panels just behind the trailing edge. Because of singular behavior of the doublet induced velocity, it forces continuity in potential and vorticity jumps at the trailing edge and finally no pressure jump at trailing edge. Direct boundary conditions (intensity of the mass outflow through the surface) can be specified for body panels. Finally, the integral equation (3) is replaced by a system of linear equations in the form:

$$[\mathbf{A}] \left\{ \begin{array}{c} \varphi \\ <\varphi> \end{array} \right\} = - \left\{ \begin{array}{c} 0 \\ \mathbf{V}_\infty \cdot \mathbf{N}_K \end{array} \right\} - [\mathbf{B}] \left\{ \dot{m}_S / \rho_\infty - \mathbf{V}_\infty \cdot \mathbf{n} \right\} . \quad (4)$$

It is solved to find the perturbation velocity potential on the surface and the jump of potential across the wake. Velocity distribution on the surface is obtained by numerical differentiation of the perturbation potential and adding the free stream contribution. In the local panel coordinate system tangent velocity components are:

$$\begin{aligned} V_t &= \partial \varphi / \partial t + \mathbf{V}_\infty \cdot \mathbf{t} \\ V_s &= \partial \varphi / \partial s + \mathbf{V}_\infty \cdot \mathbf{s} . \end{aligned} \quad (5)$$

Such a formulation is similar to many widely applied panel methods (Morino type methods). They are characterized by accuracy and important advantages in numerical implementation [4,5]. Higher order singularity distributions preserve high accuracy to computing cost ratio.

Assuming small deflection of plane flaps or control surfaces it is possible to simulate their effect by modifying the Neumann boundary conditions [6,7]. In the present method this problem is solved by specifying an alternative (modified) configuration geometry with deflected control surfaces and flaps. Assuming that AIC matrices are not so strongly affected by small geometry alterations, only **t**,**s** and **n** vectors in (4) and (5) are taken from the new geometry. The final aerodynamic coefficients are calculated by integrating pressure distribution on the modified configuration surface. It was found that the resulting pressure distributions, load distributions and aerodynamic coefficients are reasonably accurate (especially for relatively thick wings) - without additional cost for geometry preparation, AIC matrices assemblage and linear equations system solution - see Fig.1,2.

The method allows to prepare the AIC matrices for the symmetrical or both symmetrical and antisymmetrical cases. The second allows to find a solution for side-slip flow conditions and symmetrical (with respect to the x-z plane) geometry. Nonsymmetrical alternative configuration can be specified - e.g. ailerons deflection).
The above method of inviscid flow analysis is coupled with a 2-dimensional attached boundary layer code [8], allows to find boundary layer parameters on the lifting surfaces strip by strip and include weak viscous-inviscid interaction through transpiration velocity.

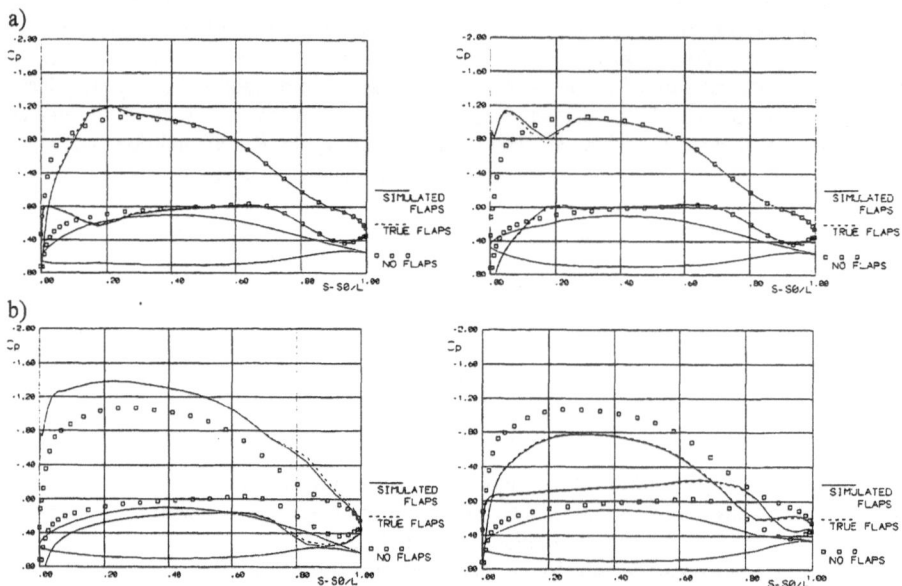

Fig. 1   Pressure distribution on the wing - without flaps (symbols), with
deflected flaps (dashed line) and with simulation of the flap deflection
(solid line):  a) leading edge flaps -/+8°, b) trailing edge flaps +/-8°

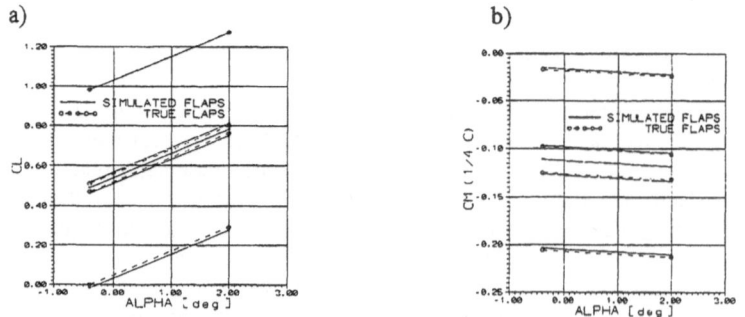

Fig. 2   Aerodynamic coefficients for the wing with true (solid line) and simulated
(dashed line) flap deflection

58

## Solution of the design problem

The most general methods of aerodynamic design are based on numerical optimization technique. In such cases it is easy to extend the problem of aerodynamic design to multi-objective optimization including aerodynamic as well as other (geometrical, structural etc.) objectives and constrains. The geometry of the target aircraft configuration is expressed as the sum of the initial geometry and a linear combination of basic geometry alterations:

$$\text{GEOMETRY} = \text{INITIAL\_GEOMETRY} + \sum_{N=1}^{ND} X_i \cdot \textit{i-th} \text{ BASIC\_SHAPE}. \quad (6)$$

The basic design shapes can be of a global type (e.g. modifications of the wing section, twist, fuselage shape) or related to the specified design points (e.g. modification of the angle of attack , control surface or flap deflection - different at various design points).

The design problem reduces thus to finding the vector of the design variables $X$ which minimize the objective function and satisfy additional constrains. In the present method the objective function consists of a number of terms:

$E_{P\_n}$    - norm of the deviation of pressure distribution from the target at n-th design point,
$E_C$    - penalty function due to violation of the allowed geometrical parameters,
$E_R$    - term enforcing smooth modification shape (regularizing term),
$C_{D\_n}$    - induced drag at a given design angle of attack,

which come into the final objective function with user-specified weights. The requested values of lift and moment coefficients can be enforced using Lagrange multipliers technique. Finally, the objective function has the form:

$$E = \sum_{n=1}^{NDP} WP_n \bullet E_{P\_n} + WC \bullet E_c + WR \bullet E_R +$$

$$\sum_{n=1}^{NDP} WD_n \bullet (C_D)_n + \sum_{n=1}^{NDP} \lambda_n^L \bullet (C_L - C_L^D)_n + \sum_{n=1}^{NDP} \lambda_n^M \bullet (C_M - C_M^D)_n \quad (7)$$

where: NDP    - number of design points (design angles of attack),
       $C_L^D$    - design lift coefficient at the n-th design point,
       $C_M^D$    - design moment coefficient at the n-th design point.

Usually the main objective is to minimize the norm of deviation of pressure distribution from the target for all the design angles of attack and configurations:

$$E_{P\_n} = \sum_{k=1}^{NPn} WP_k \bullet (V_k - V_k^D)^2 \quad (8)$$

where: NP    - number of points with prescribed velocity (or pressure),
       WP    - weight factor for the k-th panel point,
       $V^D$    - design velocity value (corresponds to the design pressure),
       V    - actual velocity at the k-th point.

Besides aerodynamic requirements, in the real design we must also take into account other requirements of geometrical or structural character [9]. Sometimes the designed geometry is completely unacceptable, e.g. negative thickness appears. Such problems often occur in designing via interference effects (e.g. designing for specified wing pressure by changing the fuselage shape). The fuselage shape, if pressure on it is not specified, is not controlled by the solution. In order to enforce the required geometrical parameters or off-design aerodynamic requirements (see [1]), certain geometrical constrains must be fulfilled. In the present method geometrical constrains are introduced via penalty function due to violation of the allowed geometrical parameters:

$$E_C = \sum_{m=1}^{NC} WC_m \cdot (\Delta H_m)^2 \qquad (9)$$

where:    NC    - number of geometrical constrains,
          $WC_m$  - weight factor for the m-th constrain,
          $\Delta H$    - measure of the violation of the geometrical constrain.

H can control a very general set of geometrical parameters: the width of the body, twist and thickness of the wing, the angle of trailing edge, etc.

As mentioned above the designed geometry is sometimes irregular. Usually it depends on the form of the specified basic geometry alterations (basic shapes). If displacements of single network point are specified as the basic design shapes, then the so called "wavy wall" and "four corner" design instabilities may occur. Sometimes the final geometry is irregular even for smooth basic shapes - especially if designing is performed via interference effects or if the (arbitrarily) specified design pressure is far from possible (no such problems have been observed in the case when geometry reconstructed from known pressure distribution). In order to remove (or just reduce) the problem we can add a new term approximating the integral curvature of the design shape along a network line:

$$E_R = \sum_{k=1}^{NR} \frac{\partial^2 h}{\partial s^2} \sim \int_{s\_ini}^{s\_end} (h''(s))^2 \, ds \qquad (10)$$

where: $h$ is the normal projection of the surface displacement along the network line.

Higher derivatives of $h$ can be also useful. It was found that the above term is sometimes a remedy which improves and simplifies the design process.

The induced drag coefficient is calculated in the Treftz plane using a far-field expression. Such a formulation can be sometimes inconsistent with the near field formulation [10] - especially for configurations with complex wake geometry. In the present method the wake geometry in the Treftz plane has the same shape as the projection of the trailing edge on this plane. It was found that for dense paneling the induced drag computed in the far field is close to the near field value. The drag increments due to modifications of the spanwise load distribution are probably much more accurately predicted then the total drag value and acceptable in optimization.

The presented panel method formulation allows to specify different (alternative) configurations for different design points. Also the angle of flap deflection for a given configuration can be specified as a design variable. Such a design variable is active only at one design point. Finally, this procedure allows to find the best combination of geometry and flaps deflections simultaneously.

The direct utilization of the panel method in the optimization routine to find the objective function is not a proper way because of the cost and computing time. In order to establish an effective design method two important problems must be solved:

- efficient and accurate method of flow analysis for perturbed geometry - to perform sensitivity analysis and directional minimization,
- specification of efficient and easy to prepare basic geometry modifications.

Because the inverse problem is nonlinear in boundary conditions, it must be solved iteratively. In the present method the flow analysis during optimization process is performed using perturbation method about an initially specified geometry. The pressure distribution resulting from such an analysis is treated as the "expected" one for the modified geometry. After each optimization cycle, a new approximation to the optimal geometry is calculated and flow analysis is performed to find the correct pressure distributions. The new geometry is assumed to be initial one for the next design iteration. The process is repeated until required convergence rate is reached. The block diagram of the method is shown in Fig. 3.

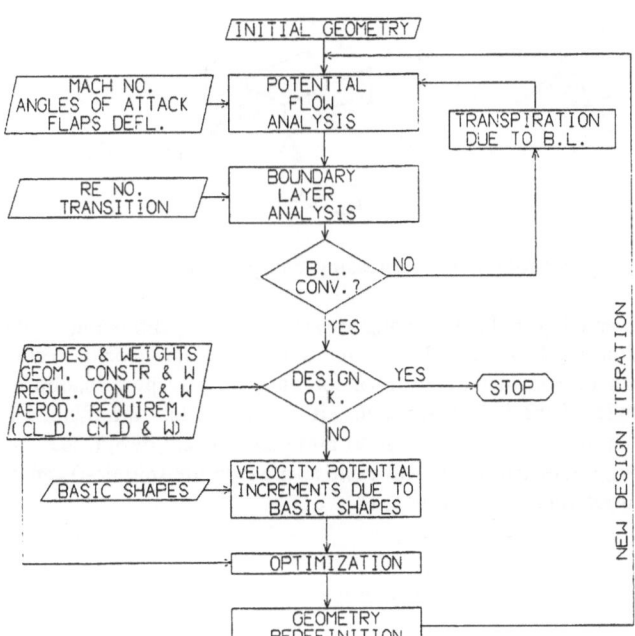

Fig.3 Block diagram
of the design
panel method

# Flow analysis on perturbed geometry

The flow analysis for a perturbed geometry can be performed in many different ways. It is crucial for the accuracy and convergence rate of the optimization process. The common method for the flow analysis mentioned above is first order expansion of the aerodynamic influence coefficients and Neumann boundary conditions with respect to the design variables [11]. The method used in the present computations is based on the transpiration concept which leads to modification of the direct Neumann boundary conditions on the initial configuration. The local modification of the geometry is modeled by mass outflow through the surface which shifts the stream surface from the initial location (the surface of the initial configuration) to the desired position. The mass flux which shifts the stream surface over the distance $h$ normal to the surface is:

$$w_{TR} = \frac{1}{\rho_\infty} \left( \frac{\partial(\rho U h)}{\partial \xi} + \frac{\partial(\rho V h)}{\partial \eta} \right). \tag{8}$$

The mean value of the transpiration through the panel area can be found by mass flux balance in the volume enclosed by the initial body surface and the target stream surface - see Fig. 4.

$$\overset{\circ}{m}_{TR} = \overset{\circ}{m}_1 + \overset{\circ}{m}_2 + \overset{\circ}{m}_3 + \overset{\circ}{m}_4 \qquad \qquad \overset{\circ}{w}_{TR} = \overset{\circ}{m}_{TR} / A_{PANEL}$$

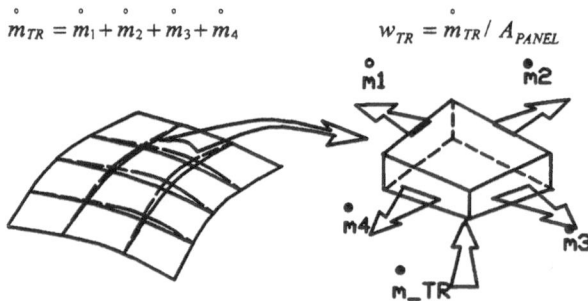

Fig. 4 Mass flux balance in the control volume expressing transpiration

For more global basic shapes (flap deflection, wing twist) a direct modification of the Neumann boundary conditions (equivalent to the rotating the free stream velocity vector) is applied. Although it can be shown that treatment of the geometry perturbation by transpiration does not include all the first order terms, it was found that the method is accurate enough and very cheap. Finally, it allows to build an efficient design method. The increments of the velocity potential due to i-th basic shape (i-th transpiration) can be found from the system of linear equations similar to (4):

$$[\mathbf{A}] \left\{ \frac{\delta\varphi}{\delta\langle\varphi\rangle} \right\}_{(i)} = -[\mathbf{B}] \{w_{TR}\}_{(i)}. \tag{9}$$

The potential on the new surface is expressed as the sum of the initial velocity potential and the linear combination (as in Eq. 6) of the potential increments due to basic design shapes. Velocity distribution is calculated in the same way as on the initial configuration (Eq. 5), using the new potential and new tangent vectors. It is easy to analytically express the first and second derivatives of the velocity components with respect to the design variables. It is useful in the optimization procedure to find the gradient and hesian of the objective function as well as lift and moment coefficients.

### Specification of basic design shapes

The geometrical problem which appears in the preparation process of the basic design shapes is sometimes even more difficult comparing to the aerodynamic one (this is especially for the user). On the one side we want to minimize the number of basic design variables, on the other it is necessary to assure a very general form of the final configuration - in order to solve the task efficiently. The problem is especially difficult in the case of a 3-dimensional complex configuration: often the basic design shapes must take into account modification of the all surface networks including modification of the intersection lines and wake surface behind the configuration. This really necessitates preparation of the new networks for the modified geometry. If the geometry of a configuration is defined parametrically, then a such modification can be done straightforwardly (if the network topology is not changed). But for more complicated configurations such a procedure is not easy. Sometimes it is better to find the modifications of the network geometry related to simplified geometry, e.g. with analytically expressed cross-sections instead of the true ones. By changing the parameters in the analytical expression of the cross-sections, it is easy to modify the entire geometry, generate the new networks and finally to find the corresponding displacement of network points. The starting geometry for flow analysis is prepared much more carefully using actually the best section shapes. Such a procedure was used in the last example presented later.

### Computer code

The described method was coded in MS FORTRAN PowerStation and implemented on PC-Computers. The software was prepared as a package. All basic elements of the method are separate programs. There are 14 of them, including two solvers of the system of linear equations: an iterative one and the Crout LU decomposition, a post-processing program and the off-body velocity evaluation. It is possible to use up to about 6000 body panels (on the half of the configuration), 240 basic design shapes and 3 design points. Flow analysis on Pentium 133MHz (HX) and with 2400 body panels on half of the configuration takes about 20' if the Crout decomposition method is used and 10' with the iterative method. For 4000 body panels the computing time is about 80' and 30', respectively. The entire 2-point design iteration using 2400 body panels and 200 basic design shapes took about 30'.

## Results

Sailplane design. A high performance glider must have a high L/D ratio at various flight conditions: from low CL at high speed to high CL at low speed. In order to extend the CL range of low drag laminar flow conditions, the wing sections can be equipped with the trailing edge flaps. In the case of the entire glider configuration, the expected characteristics of high performance profile decline due to 3-dimensional interference effects. The range of lift coefficient corresponding to low drag of the laminar flow profile is significantly reduced for the wing-fuselage combination. This is mainly due to the so called fuselage cross-flow which leads to locally (near the fuselage) higher angle of attack (compared to the geometrical one) at positive $\alpha$ and lower at a negative one. The lift coefficient can be even lower due to the effect of "non-lifting" fuselage. At high CL the induced drag is important and winglets can be used to reduce it. On the other hand, such a solution can lead to the increase of drag at low CL due to winglet off-design conditions and a higher friction drag. The mentioned below example presents the study case of the application the hybrid method: inverse design - optimization technique to improve performance characteristics. No effect of laminar separation bubbles is incorporated. Transition is predicted by $e^9$ method, or it is assumed to happen at a laminar separation point. Two configurations were compared in order to evaluate the possible effect:

1. flat wing - fuselage - horizontal tail configuration,
2. wing-winglet - fuselage - horizontal tail configuration,

The second configuration has the total wetted area increased by about 0.7 %, but exactly the same reference area. Both configurations have initially constant section with 2-dimensional optimum of flap deflection, without twist. Wing planform has a nearly optimum chord distribution and a straight trailing edge. Aerodynamic characteristics (without fuselage drag) in trim were evaluated at two flight conditions:

- high speed: CL = 0.16, Re = 3.0 mln.
- low speed: CL = 0.87, Re = 1.2 mln.

The designing was performed in order to eliminate adverse interference effects (by specifying the design pressure distribution) and to minimize induced drag at both design flight conditions. The first configuration was divided into 2430 body panels and 192 basic design shapes were utilized: wing section shape and twist modifications at 6 stations, fuselage modifications (independent for upper and lower parts), flaps deflection (as three segments) and angle of attack. The second configuration was divided into 2560 panels and 222 basic design shapes were specified. In addition to the above mentioned basic shapes, the winglet section shape and a small winglet flap (in order to adapt the winglet to various CL conditions) were modified. The isobar pattern on the first configuration before designing is shown in Fig. 5. Similar isobar pattern is on the second starting configuration, with additional problem in a concave part of wing-winglet junction (higher negative pressure and pressure gradient). After designing the isobar pattern is significantly improved, giving nearly constant pressure distribution at both configurations. The design pressure on winglets is different from its value on the basic part of the wing: the negative pressure is not so low in order to improve the boundary layer flow at locally very low Reynolds number. The isobar pattern on the second configuration after designing is shown in Fig. 6 and the pressure distribution at both angles of attack in the section near the fuselage in Fig. 7. Next figures present distribution of the local lift coefficient and expected location of transition on the lower surface at low CL and on the upper surface at higher CL. The predicted improvement in the sum of the induced drag and the friction drag on the lifting surfaces is significant. It can be expected that the careful designing of the configuration with winglets will reduce total drag at low CL by about 3 drag counts as compared to the initial configuration without winglets (due to removing fuselage interference and despite the additional friction drag of the winglet) and nearly 6 drag counts at higher CL (induced drag reduced by 5% compared to the flat wing). It means performance improvement of about 4% at high speed and nearly 3% at low speed (including fuselage drag).

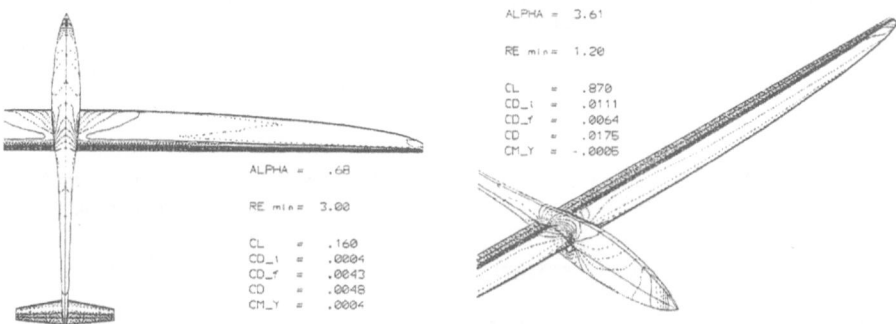

Fig.5 Isobar pattern on the initial glider configuration with a flat wing (low and high CL).

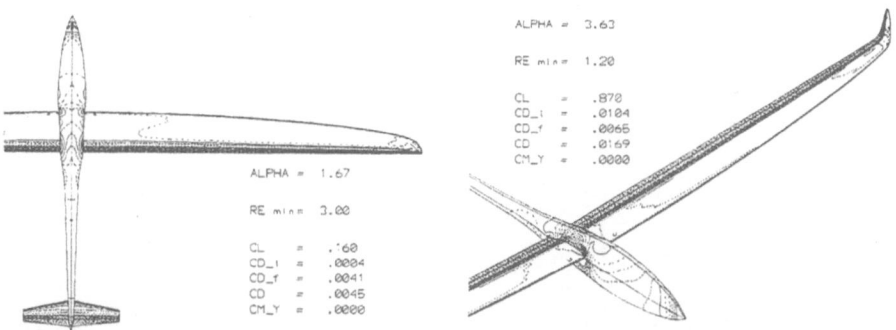

Fig.6 Isobar pattern on the designed glider configuration with winglets.

Apart from the above presented adverse interference effects of the inviscid origin there are also viscous interference effects [12]: the laminar boundary layer on the fuselage forebody turns turbulent and separates as it approaches the wing root stagnation pressure. Thus it modifies the wing boundary layer flow near the wing-fuselage junction. It can cause separation on the wing section designed for laminar flow.

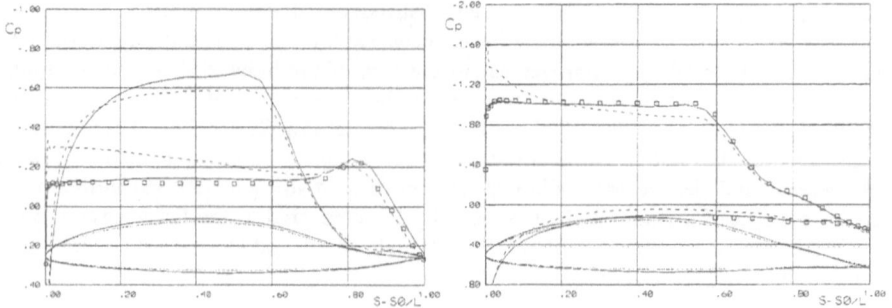

Fig. 7 Wing section pressure distribution near the fuselage at both design conditions:
initial (dashed line), final (solid line) and target (symbols)

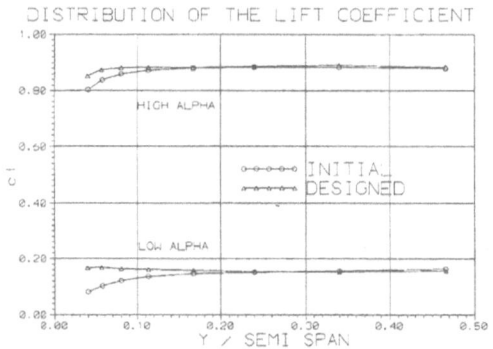

Fig. 8 Spanwise distribution of the
lift coefficient before and
after designing

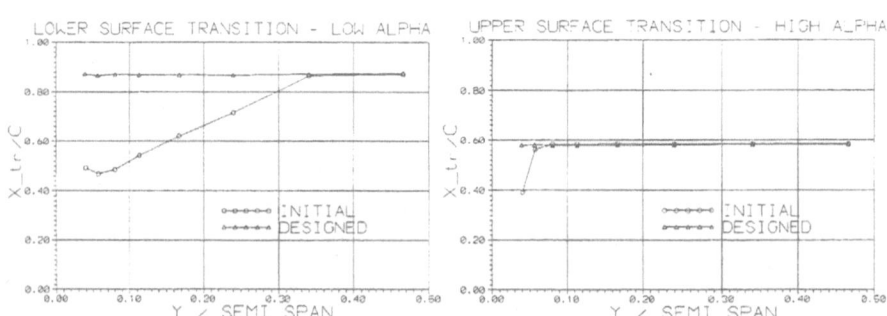

Fig. 9 Predicted transition location at both design points before and after designing

The proper treatment of this effects requires modification of the design pressure in this region. Fig. 10 presents an interesting nonplanar glider configuration designed and tested at Delft University of Technology and actually manufacturing in Germany. It has nearly eliminated the cross-flow effect [13]. A two-point design was performed, corresponding to lift coefficient respectively 1.0 and 0.42 (without flaps). Variation of the design pressure is visible on the wing. Wing sections were carefully designed with 2-dimensional method [8] and tested in wind tunnel. Similar procedure making use of the above described panel method was applied at Delft University of Technology for wing-fuselage design of the new gliders, including actually flying ASW 27 - with a sophisticated wing-fuselage junction.

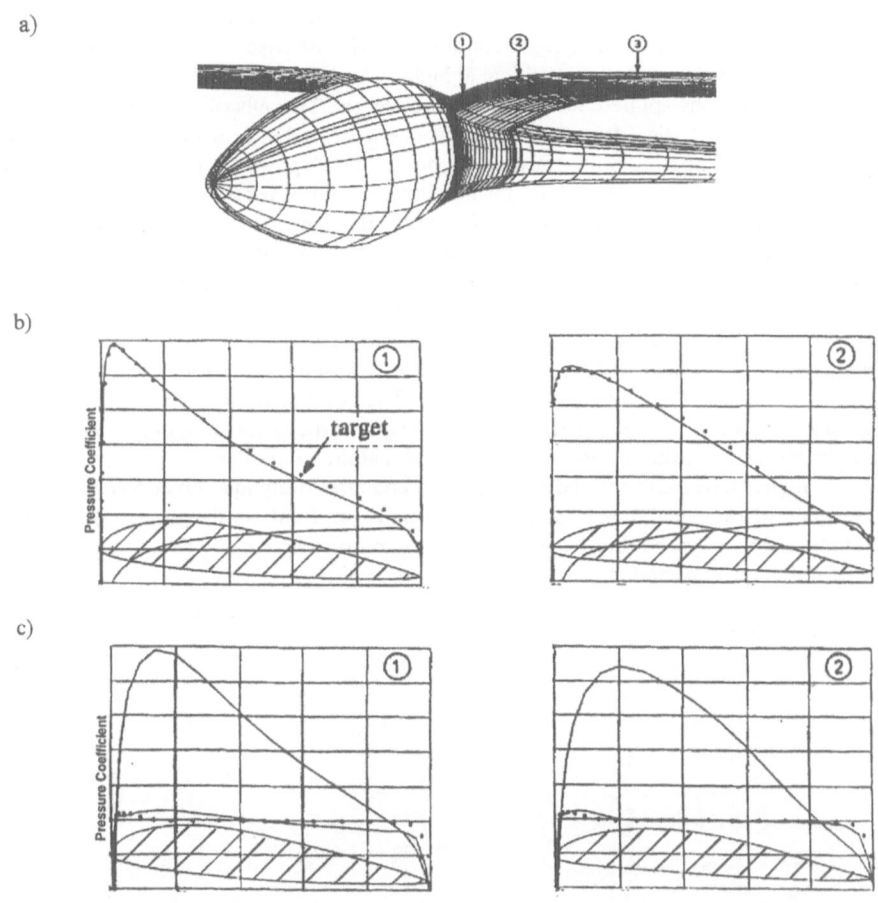

Fig. 10 Nonplanar wing-fuselage glider 2-point design overview (a),
high CL (b) and low CL (c) conditions.

Design study of the 3-surface aircraft configuration. The above panel method was applied to perform a preliminary design study of an original ground-attack aircraft configuration. The design objectives are rather unusual. The aircraft should be small (overall length of about 10.5 m) and light , powered by two high bypass turbofans with relatively large dimensions. The aircraft dimensions eliminate the possibility to locate the engines in the fuselage. In order to find room for appreciable amount of fuel, armament and equipment a large internal volume near the center of gravity is required. Also the thickness of the central section of the wing should be big. The considered configuration (Fig. 11) has three-surface layout with a low aspect ratio (AR~3) main wing, residual fuselage and engines mounted behind the load-bearing structure with inlets over the wing. The entire configuration has an extremely closely coupled layout with strong aerodynamic interference. The forward wing is provided mainly as the slat for the central part of the main wing in order to unload it. Finally, it should eliminate the high flow disturbances in the inlets region at higher angles of attack, and reduce the value of the negative pressure below the critical at higher Mach numbers and angles of attack (for thick wing sections applied at this part of the wing). The subcritical panel method was used for the supercritical flow conditions ($Ma_\infty = 0.80$) in hope, that a designed geometry will be the good starting point for the next design development. The design pressure for the outer part of the main wing and for the forward wing is taken from the 2-dimensional wing sections designed for transonic flow (via the concept of "equivalent subcritical pressure distribution" [14]). In the central part of the wing and nacelles a completely subcritical pressure distribution was specified. As before, the pressure distribution and induced drag minimization at two angles of attack (low and higher CL) and at trim conditions were the design objectives. The starting geometry of the configuration was prepared by an initial evaluation the thickness/volume effects and using the lifting surface theory for mean surfaces evaluation. The configuration was divided into about 3900 body panels. Totally 141 basic shapes were utilized for designing. The Fig. 12 shows the isobar pattern and calculated aerodynamic coefficients at the second design point (higher CL) for the initial configuration and after designing. Over 5% reduction of the induced drag and better pressure (especially in front of the inlets) is observed. Additionally low speed, high angle of attack conditions were examined, computationally confirming the expectations.

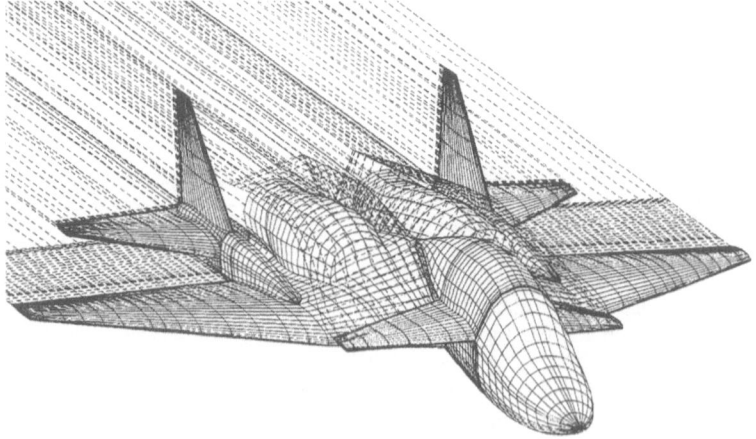

Fig. 11  General layout of the 3-surface aircraft configuration

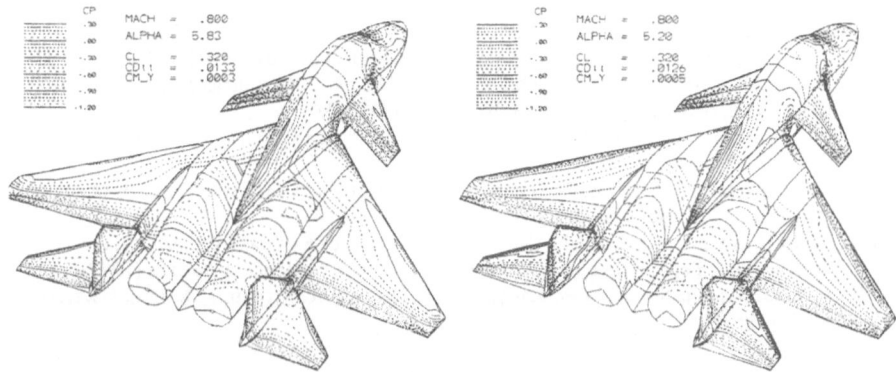

Fig. 12  Isobar pattern and induced drag on the initial and designed configuration.

## Concluding remarks

In the paper a very general method for subcritical aerodynamic design has been presented. It allows to design real, complex configurations with simply connected maneuver flaps, including the multi-point inverse design and induced drag minimization. The major limitation is the lack of possibility to include the planform shape as a design parameter, restrictions to linear compressibility effects and weak viscous-inviscid interaction. One of the main advantages of the method is the possibility to include the interference effects in designing, e.g. applying fuselage shape modifications in order to produce required wing flow properties.

## References

[1]  K. Kubrynski, Design of 3-Dimensional Complex Airplane Configurations with Specified Pressure Distribution via Optimization, Proceedings of ICIDES-III Conference, G.S. Dulikravich, Washington D.C. (1991).

[2]  K. Kubrynski, Two-Point Optimization of Complete Three-Dimensional Airplane Configuration, AIAA 10'th Applied Aerodynamics Conference, AIAA 92-2618 (1992).

[3]  G. N. Ward, Linearized Theory of Steady High-Speed Flow, Cambridge University Press (1995).

[4]  D. R. Bristow, G. G. Grose, Modification of the Douglass Neumann Program to Improve the Efficiency of Predicting Component Interference and High Lift Characteristics, NASA CR-3020 (1978).

[5]  F. T. Johnson, A General Panel Method for the Analysis and Design of Arbitrary Configurations in Incompressible Flow, NASA CR-3079 (1980).

[6]  S. Heiss, L. Fornasier, Analysis of a Fighter Type Aircraft Configuration with the HISSS Panel Method at Subsonic and Supersonic Speeds, Z. Flugwiss. Weltraumforsch. 12 (1998), pp.224-232.

[7]  S. A. Moyer, PAN AIR Analysis of Simply Connected Control Surface Deflections, J. Aircraft, Vol.30 No. 1 (1993).

[8]  M. Drela, XFOIL: An Analysis and Design System for Low Reynolds Number Airfoils, in Low Reynolds Number Aerodynamics, Ed. T. J. Mueller , Lecture Notes in Eng.  54 (1989).

[9]  J. M. J. Fray, J. W. Slooff, A Constrained Inverse Method for the Aerodynamic Design of Thick Wings with Given Pressure Distribution in Subsonic Flow,  AGARD CP-285 (1980).

[10] I.Kroo, S.C.Smith, The Computation of Induced Drag with Nonplanar and Deformed Wakes, SAE Technical Paper Series, No. 901933 (1990).

[11] D. R. Bristow, J. D. Hawk, Subsonic Panel Method for the Efficient Analysis of Multiple Geometry Perturbations, NASA CR-3528 (1982).

[12] L.M.M.Boermans, D.C.Terleth, Wind Tunnel Tests of Eight Sailplane Wing-Fuselage Combinations, Technical Soaring, Vol. VIII, No. 3  (1984).

[13] L.M.M.Boermans, K.Kubrynski, Aerodynamic Design of Sailplane Wing-Fuselage Combinations, Paper presented at  23 OSTIV Congress, Botlange, Sweeden (1993).

[14] J.W.Sloof, Application of Computational Procedures in Aerodynamic Design, AGARD, R-712 (1983).

# Parametric Airfoils and Wings

Helmut Sobieczky

DLR German Aerospace Research Establishment
Bunsenstr. 10, D-37073 Göttingen ,Germany
e-mail: Helmut.Sobieczky@dlr.de

## Summary

Explicit mathematical functions are used for 2D curve definition for airfoil design. Flowphenomena-oriented parameters control geometrical and aerodynamic properties. Airfoil shapes are blended with known analytical section formulae. Generic variable camber wing sections and multicomponent airfoils are generated. For 3D wing definition all parameters are made functions of a third spanwise coordinate. High lift systems are defined kinematically by modelled track gear geometries, translation and rotation in 3D space. Examples for parameter variation in numerical optimization, mechanical adaptation and for unsteady coupling of flow and configuration are presented.

## Introduction

Airfoil and wing design methodologies have made large steps forward through the availability of rapid computational tools which allow for specification of goals in aerodynamic performance. These goals are mainly to increase a measure of efficiency, like the ratio of lift over drag, or, in the higher speed regimes, its product with flight Mach number. The need for increased lift at higher flight speed, with drag kept low, has led to the development of knowledge bases for aerodynamic design: The art of shaping lift generating devices like aircraft wings is based on geometric, mechanical and fluid dynamic modelling, carried out with the help of mathematical tools on rapid computers. Given a designer's refined knowledge about the occurring flow phenomena, his goal may be to obtain certain pressure distributions on wing surfaces: This may be reached by inverse approaches with a shape resulting from the effort, or by applying optimization strategies to drive results toward ideal values.

With such methods we have refined tools available for extending our practical knowledge how the geometries of airfoils and wings are related to pressure distributions and aerodynamic performance. Certain details of desirable pressure distributions require a modelling of details in the boundary condition, usually a special feature of the curvature distribution. This is true especially in the transonic flow regime, where favorable as well as undesirable aerodynamic phenomena are correctly modelled by certain weak or strong singularities in the local mathematical flow structure including the flow boundary. Numerical optimization methods iteratively adjusting the resulting 2D or 3D shapes usually employ smoothing algorithms based on polynomials, splines and similar algebraic functions. These functions may be ignoring local properties of the shape being compatible to the inverse input, while they should accomodate the results from analytical inverse methodology using hodograph formulations of the governing equations. Hodograph-type methods, though not practical tools, have led to a deeper understanding about the relations between surface geometry and the structure of recompression shocks. These methods are most usefully applied to designing nearly shock-free airfoils and wings with favorable off-design behavior. Understanding the resulting refined shapes and modelling them in a direct approach with a suitable geometry generator is a continuing challenge for more complex 3D configurations like complete aircraft, turbomachinery components and models for interdisciplinary design.

The present contribution is aimed at using explicit mathematical functions with a set of free parameters to define wing surfaces of practical interest for realistic aircraft applications, with a potential to arrive at optimum values of objective functions like aerodynamic efficiency, with a minimum of parameters having to be varied, because these parameters are defined from application of the fluid mechanic and gasdynamic knowledge base or prescribed by modelling kinematic models of a mechanical adaptation device.

## Geometry generator

In the series of Notes on Numerical Fluid Mechanics the author has had the chance to present concepts, tools and examples of shape definition for aerodynamic components, with a strong emphasis on using mathematical functions which are drawn from analytical modelling of flow phenomena as they occur in the transonic regime. The need for reduction of shock losses has sparked an inverse procedure to find shock-free airfoils and wings, with the additional option to adapt wing geometries to varying operating conditions [1]. The increased need for creating test cases for numerical flow simulation (CFD), along with the requirements for precise definition of boundary conditions has then inspired the presentation of a wing within a transonic wind tunnel, with all boundaries including the tunnel and the inlet and exit flow conditions given [2], to be simulated and compared with experiments [3]. Later, the mathematical tools for defining such boundary conditions were further developed to model real aircraft components: wings, fuselages, propulsion components and their integration to complete configurations [4]. Since then, various applications have been studied and more recent refinements led to several versions of "geometry preprocessor software tools". These support modern developments in a multidisciplinary design environment for aerospace components and not restricted to aerodynamic optimization.

Aircraft wings are the primary subject to optimization efforts, progress in aerodynamic design methodology is mostly influenced by new ideas to improve the lift-generating devices. Airfoils are the basic elements of wing geometry, they determine a large share of wing flow phenomena though they are just two-dimensional (2D) sections of the physical wing surface. Well-known aspects of wing theory are the reason for options of such idealization, with a large accumulated knowledge base resulting for 2D airfoil theory. It has, therefore, been well founded to use airfoil shapes with documented performance results from wind tunnel tests for the design of wing shapes. These airfoils are usually contained in published or proprietary data bases, we use them as dense data sets to describe the sections of wings with planform, twist and dihedral given by analytical model functions. Properties relevant for flow quality, for instance curvature, of these latter functions are simple and easily controlled by parameters while the airfoil input data are to be spline-interpolated to obtain a required distribution of surface data. With all the experience gained by using our shape-generating tools and updating them with recent developments in designing high speed flow examples, an effort is made to generate 2D wing sections in the same way the 3D shape parameters are already defined. Suitable functions should replace the hitherto required airfoil data sets. The goal is to propose functions with a minimum set of input parameters for shape variation, function structure and their parameters chosen to address special aerodynamic or fluid mechanic phenomena. This desirably relatively small number of control parameters will then effectively support optimization procedures.

# Airfoil functions

With airfoil theory and airfoil data bases being well established components of applied aerody-namics on the ground of lifting wing theory, it is necessary to allow for using such data as a direct input in any wing geometry definition program. This fact was the motivation to provide spline interpolation for such given airfoil data in a first version of our geometry code, which has been described in various papers and publications. Recently these developments have been summarized in [5], here we focus on continuing this activity in the area of describing airfoils with more a sophisticated method than providing a set of spline supports.

Functions to describe airfoil sections are known for many applications, like the NACA 4 and 5 digit airfoils and other standard sections. Aircraft and turbomachinery industry have developed their own mathematical tools to create specific wing and blade sections, suitably allowing par-ametric variation within certain boundaries. We define such functions for airfoil coordinates in coordinates X, Z non-dimensionalized with wing chord therefore quite generally

$$Z = F_j(p, X)$$

with $p = (p_1, p_2, ..., p_k)$ a parameter vector with k components and $F_j$ a special function using these parameters in a way determined by a switch j. The goal is to try to keep the number k of needed parameters as low as possible while controlling the important aerodynamic features ef-fectively.

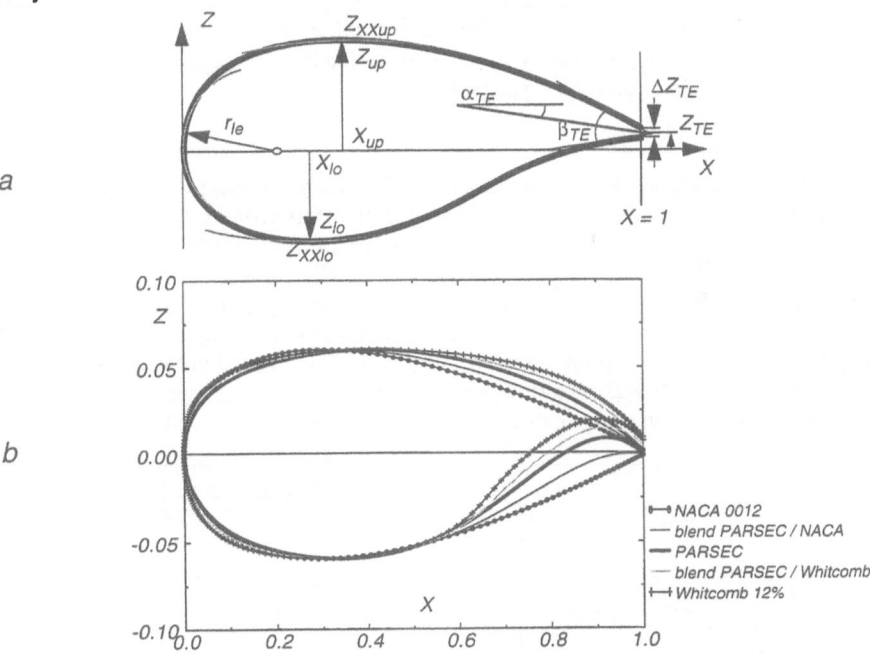

Fig. 1: "PARSEC" airfoil geometry defined by 11 basic parameters: leading edge ra-dius, upper and lower crest location including curvature there, trailing edge coordi-nate (at X = 1), thickness, direction and wedge angle, (a).
Example: Variations of PARSEC airfoil by blending with NACA or Whitcomb airfoil (b)

Figure 1 illustrates 11 basic parameters for an airfoil family "PARSEC" which we found quite useful for applications. There is strong control over curvature by prescribing leading edge radius and upper and lower crest curvatures. Similar to 4Digit NACA series we choose a polynomial, though of a higher (6th) order:

$$Z_{PARSEC} = \sum_{n=1}^{6} a_n(p) \cdot X^{n-1/2}$$

for upper and lower surface independently, the coefficients $a_n$ determined from the given geometric parameters as illustrated in Fig.1. Comparison with other new or well known airfoil generator functions is made possible by including those functions in the software, a combination of individual features is then straightforward:

*Blending of different airfoil generator functions*

With an additional blending parameter $p_{mix}$, some known airfoils are included in this geometry tool as basic default functions, like NACA series airfoils as coded by Ladson [6]and Whitcomb's supercritical wing sections as coded by Eberle [7]. These known sections require input of a subgroup of the above 11 basic parameters and they can be blended in with the more refined geometries.
Figure 1b shows an example of an airfoil series whith the 11 basic parameters kept fixed and using only the blending parameter, resulting in two different variations of the special choice PARSEC airfoil: Blending with an NACA 4Digit section for $-1 < p_{mix} < 0$ and blending with a 12% thick Whitcomb airfoil for $0 < p_{mix} < 1$.

*Example: Transonic airfoils*

The 11 basic parameters in Fig. 1b are selected to re-model an efficient wing section from a previous study using the above mentioned spline support airfoil definition technique. A 30°swept wing with a 12% thick main section had been designed for Mach = 0.85 and Re = 40 Mill. First favorable results of CFD analysis suggested a more detailed development of this wing and its main section. The PARSEC routine is applied here by choosing the 11 basic parameters directly from analysis of the spline support section. Application of swept wing theory requires a thickening by a factor of 1.1547 resulting in the airfoil to be investigated in Mach = 0.74. Here and in the following the Drela-Zores airfoil expert system software [8] is used for fast viscous transonic flow analysis, with pressure distributions, dragrise and polars resulting, Fig. 2.

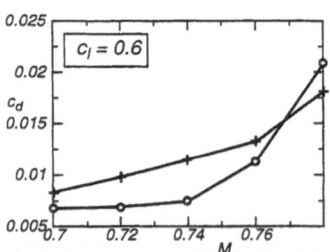 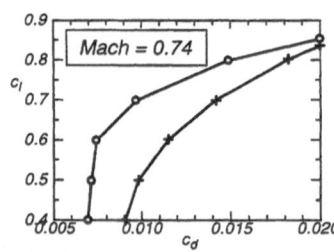

Fig. 2: Dragrise and lift/drag polar for Re = 40 Mill. transonic flow past PARSEC airfoil (symbol o) and Whitcomb airfoil (symbol +) using 6 of the 11 basic parameters. Design conditions at Mach = 0.74, $c_l$ = 0.6.

Comparative Whitcomb section results are shown, too, indicating that for the relatively low transonic design Mach number 0.74 the PARSEC airfoil seems well suited, (ratio lift/drag = 80), while the use in higher design Mach numbers a blending with a Whitcomb section - or a variation of the basic PARSEC parameters guided by the Whitcomb airfoil geometry - will be a way to optimize the wing section with a relatively modest effort.

*Parameter variations*

The parametric airfoil generator PARSEC allows for control of curvature at the nose, at the upper and at the lower crest. With these additional degrees of freedom - compared to airfoil functions without curvature control - we may vary aerodynamic performance and shift the optimum conditions to desired operation conditions. The example illustrated in Fig. 3 shows a variation of the above PARSEC airfoil by, first, only increasing the leading edge radius and, second, also decreasing the upper crest curvature, which is suggested by the analyzed curvature values for the Whitcomb airfoil. We see a shift of the drag rise toward higher Mach numbers. Other parameter variations give similar substantial changes in performance. Here we stress the observed fact that for the PARSEC airfoil model function some *single* parameter changes may already improve a given section for selected operating conditions.

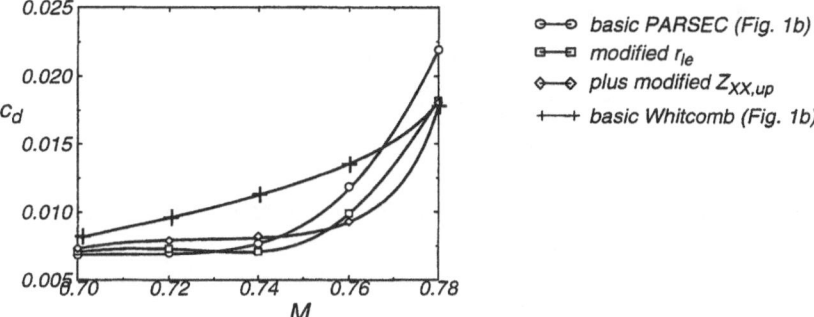

Fig. 3: *Shifting drag rise to higher Mach numbers by changing single parameters*

*Trailing edge (TE) variations*

Refined control of viscous flow parameters near the wing trailing edge may influence circulation and hence aerodynamic efficiency quite remarkably. In the past this has led to specially designed airfoil and wing TE shapes: Special solutions to the outer inviscid flow model equations were proposed to create a flow field in the vicinity of the TE which has a favorable pressure gradient on the airfoil surface. One little known theoretical base for the shaping of high performance airfoils has been presented by Garabedian [9]. Based on a complex hodograph analysis, the principle can be modelled by increased curvature only quite closely at the TE, to counteract the boundary layer's de-cambering effect.

This occurs on the upper surface more locally than on the lower surface. The practical consequence for physically relevant airfoils which are not having negative thickness or too thin TE's, is a blunt TE base, a convex upper surface contour shaping with curvature increasing toward the TE and a more evenly distributed curvature on the concave lower surface, resulting in a minimum thickness of the airfoil a few percent upstream of the TE. Such TE refinements have been studied on practical wing sections and have been termed 'Divergent Trailing Edge - DTE' airfoils [10],[11]. Modifications based on the hodograph analysis are added to the basic PARSEC shape: In the simplest case a single additional parameter $\Delta\alpha$ controls the functions added to airfoil upper and lower surface to become a DTE wing section, see Figure 4. Based on our hitherto quite limited experience with case studies, modification lengths $L_{1,2}$ range between 20 and 50 % of airfoil chord, for the exponents values we use $n = 3$ and $1.8 > \mu > 1.3$ (Garabedian's hodograph solutions suggest $\mu = 4/3$).

$$\Delta Z = \frac{L \cdot \tan\Delta\alpha}{\mu \cdot n} \cdot [\, 1 - \mu\xi^n - (1 - \xi^n)^\mu \,]$$

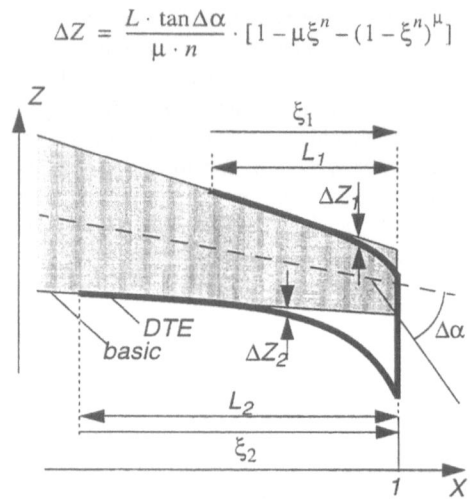

*Fig. 4: Local airfoil geometry modifications to model a divergent trailing edge.*

*Local surface bumps*

Transonic airfoils of high efficiency differ from classical low speed airfoils mainly due to their delicate curvature distribution on the upper (lifting) surface where supersonic flow conditions occur. In the case of exactly shock-free flow we observe distinct curvature maxima close to where local Mach number unity flow is found; these shape details can be understood from locally valid model solutions to the inviscid basic equations which have led to systematic design methods based on operational CFD analysis codes; a review of this concept with a more detailed discussion of the interaction of transonic gasdynamics with geometrical boundaries is given by the author in [12]. It is shown that the addition of two suitable bumps to a given conventional airfoil can convert the flow to be shock-free in high subsonic Mach numbers. This is done by a first bump near the leading edge which triggers a cluster of expansion waves, and a second one absorbs recompression waves which coalesce near the sonic recompression.

Based on this established knowledge for the design of transonic airfoils with high aerodynamic efficiency it has been found useful to influence local curvature of given airfoils in critical regions by surface bumps of varying shape and size, which can be built in an aircraft wing as a flow control device. Even reducing this effort to only adding a very small bump, extending to 2 - 3 % of chord near the location of a recompression shock, has been claimed to improve aerodynamic performance by dispersing the shock at the foot point on the airfoil, thus favorably influencing shock - boundary layer interaction [13].

$$\Delta Z = Z_m \cdot \sin(f(\xi))^{g(\xi)}$$

$$f(\xi) = a\xi + b\xi^2$$

$$g(\xi) = (P - Q\xi) \cdot (1 - (1 - c) \cdot \sin\xi)$$

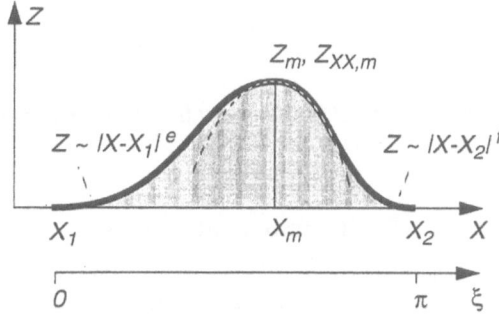

*Fig. 5: Local airfoil geometry modifications to model a bump with strong shape control.*

A function for arbitrary bumps has been added therefore to the airfoil generator program which generates bumps (Fig. 5) with strong local curvature control. Chordwise extent $(X_1, X_2)$ is defined by choice of the local variable $\xi$. Possible requirement of an unsymmetrical bump crest location $(X_m, Z_m)$ will be taken care for by coefficients a and b, crest curvature $(Z_{XXm})$ and curvature control at the bump ramps (e, f) are controlled by the coefficients P, Q and c in the equation for g(x) as can easily be verified.

*Variable camber models*

Flexible parts for a wing geometry may be used for widening the range of optimum efficiency in variable operating conditions. Structural constraints restrict the use of such parts to the areas of trailing and leading edges. For ensuring acceptable flow quality, sealed flaps at the TE and sealed slats at the leading edge (LE) maintain the smooth contour without gaps or corners. Setting up a geometrical model for realistic wings or wing sections of course depends on knowledge of the mechanical device putting the concept to reality. For rapid predesign studies, the task to geometrically model such contour variations is to define only the deflection angle $\omega$ as a single input parameter, for a given kinematics (in the simplest case the hinge point H coordinates) and fixed domain $(\Delta X)$ of some elastic surface modification. The sketch Figure 6 illustrates this for both a sealed slat and a sealed flap.

Without further specification of the elastic surface mechanics a simple analytic function provides a smooth connection between the original airfoil and its rotated nose or tail portion. For a specified mechanism solving the problem of connecting the solid parts with an elastic and sealed contour the model function needs to be adapted to the hardware data.

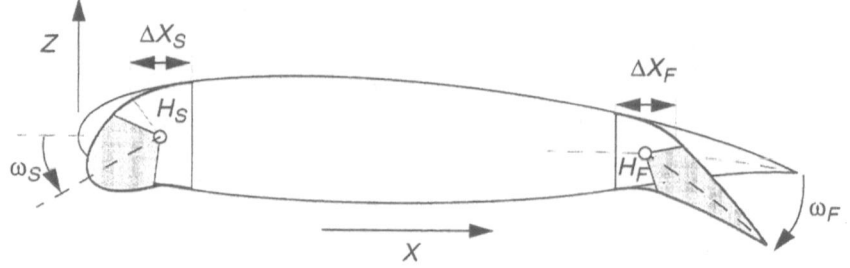

*Fig. 6: Local geometry variation at leading and trailing edges by sealed slat and flap model*

## Multicomponent airfoils

A more complex task of generic parameterized modelling is the geometry of high lift components like Fowler flaps and slats. Here we start also from a given airfoil, but we need to carve out separated lift-generating airfoils from the nose area and one or more of such sections at the rear portion of the basic airfoil. Figure 7 shows the added geometry details for a given airfoil modified to include a single slat which can be moved by a combined translation and rotation. Choice of coordinates for $C_0$, $C_1$ and $C_2$ and curvatures there define curve functions similar to the above PARSEC approach for the remaining fixed airfoil portion (or similarly at the flap nose portion).

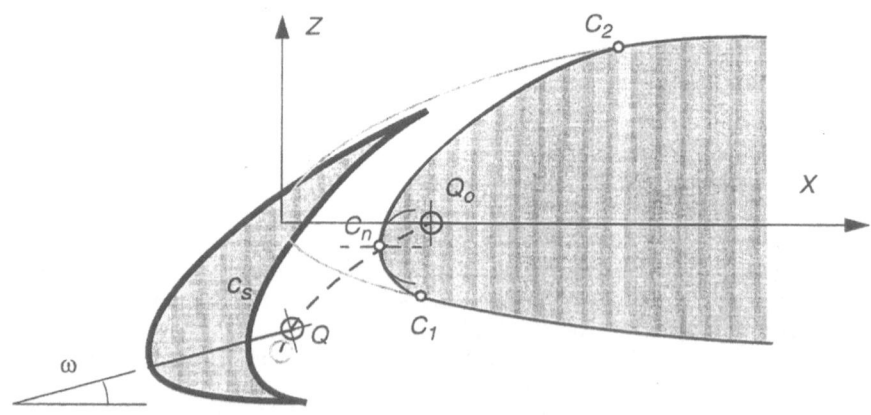

*Fig. 7: Selecting a portion of airfoil contour ($C_1C_2$) to carve a slat geometry $c_s$. Added functions for carved surface, coupled translation ($Q_oQ$) and rotation angle $\omega$.*

Figure 8 shows an airfoil with a single flap, various positions depicted. Drela's code for multi-component airfoils [8] can be used for a rapid manual optimization of the flap track to obtain high lift coefficients. An estimation of the separation bubble displacement within the flap bay, to be modelled for each flap (or slat) position, is quite helpful for these pre-design studies: with the same approach it is straightforward to model also a viscous displacement contour $c_v$ to replace the concave surface parts. More refined analysis using a Navier/Stokes solver is needed to calibrate $c_v$ for the faster analysis methods, but the parameters to do so in a flexible way may be available already.

It should be stressed that these 2D multicomponent airfoils are to be used in the cruising (retracted) configuration only for 3D applications: Swept wings will require a 3D definition of flap tracks and a shifting and rotation of the whole 3D flap or slat, which cannot be modelled for each section in a 2D fashion. The modelled retracted 2D components are the baseline for the real 3D high lift system.

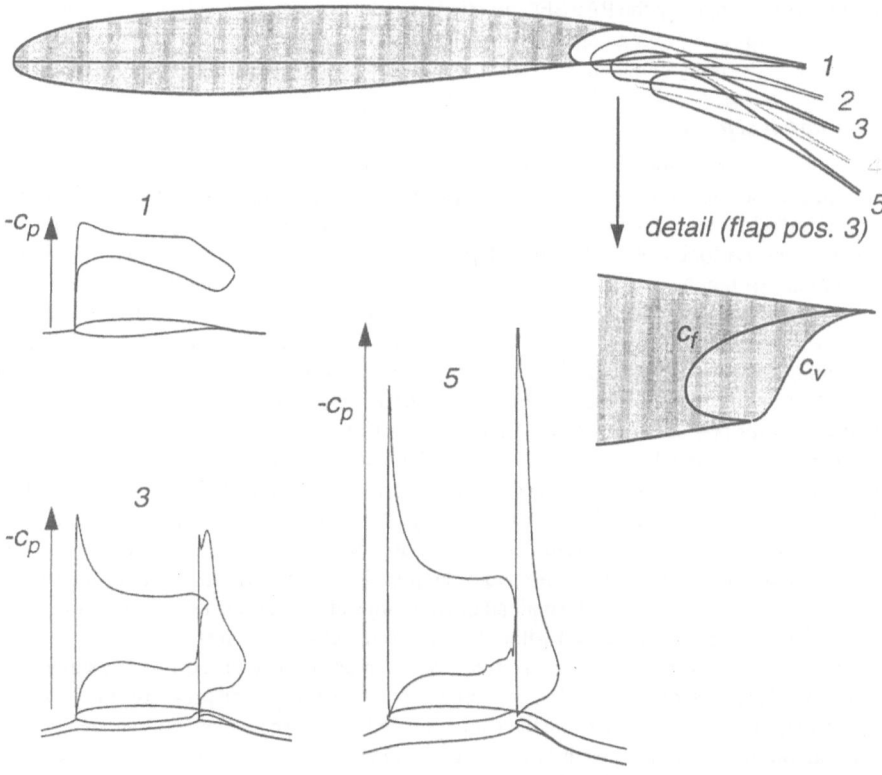

Fig. 8: Modelling flap geometry $c_f$ and viscous flow replacement contour $c_v$ for fast 2D flow high lift computation. Pressure distributions for original airfoil (1) and two flap positions (3, 5) in incompressible, inviscid flow, estimated separation in flap bay modelled.

# Wing geometries with spanwise section variation

So far our shaping of aerodynamic components is restricted to a 2D space (X, Z), which is non-dimensionalized with airfoil chord. In the following, this chord will be a function of the spanwise coordinate, $Y = y_0$, of a 3D wing, which is the independent variable to scale, shift and rotate each wing section in 3D space (x, y, z).

The flexible geometry generator for 3D wings, also laid out for curve and surface definition based on suitable parameter input [5], so far makes use of a number of airfoils as 'support sections' at given spanwise positions; blending functions defined within the resulting intervals give a section geometry at every spanwise station. For an already very precisely given wing with many support sections and small intervals, this section blending is used merely for a linear interpolation to obtain a redistributed or refined surface grid. Such approach, in principle, may lead to inaccuracies in spanwise smoothness, which does not occur from a definition with only few support sections.

Application of the analytically defined airfoils as wing sections with a smooth variation of the parameters (here, applying the PARSEC functions, the 11 basic and optionally a few parameters for surface bumps and trailing edge variations) guarantees surface smoothness to a desired degree with also just few input data and still an option for strong surface variation along span.

## *Periodic airfoil deformation*

Not yet a wing configuration, but introducing time as a third dimension defines a 3D boundary condition, see the illustration Figure 7: We applied the new parameterized shapes for the generation of airfoil systems with periodic geometry changes. So far we have studied applications to new helicopter rotor blades with shape adaptation for suppression of dynamic stall: a periodic nose drooping to a given airfoil using the geometry manipulation as illustrated for the sealed slat, Fig. 6, was applied, in certain phase with a periodically changing angle of attack. Results are obtained indicating a very favorable delay of unsteady separation in the low speed phase of the retreating rotor blade [14]. An unsteady N/S code is used with a grid conforming and moving with the varying airfoil geometry. This concept of using periodically adaptive airfoils is also applied with a refined nose curvature variation to control shock-boundary layer interaction in high angle of attack airfoil flow [15]. The study revealed the dramatic role of viscous transonic phenomena occurring in low speed aerodynamics: a small supersonic bubble forms at the leading edge at high angle of attack. Because of high curvature the recompression shock terminating this supersonic domain is quite strong. Our earlier approach to design shock-free supercritical airfoils has taught us to apply bumps for shock suppression; it seems that such approach may be successfully used very locally with small adaptive bumps at the leading edge in low speed configurations, which might be realized without excessive mechanical effort.

Software tools developed for the comparative visualization of 3D CFD results are suitably applied here for 2D unsteady flows: Color or 'zebra' isofringes for surface pressure show the onset of separation, Fig. 9. Iso-surfaces like the sonic bubble M = 1, are displayed as shaded surfaces and give an impression of the extent of the observed phenomena. Creating this 3D visualization of an unsteady 2D flow, as well as the use of video animation of such case studies and their 4D extension if a 3D unsteady process is being investigated, have a high educational value for identifying the relative importance of single parameters to be varied.

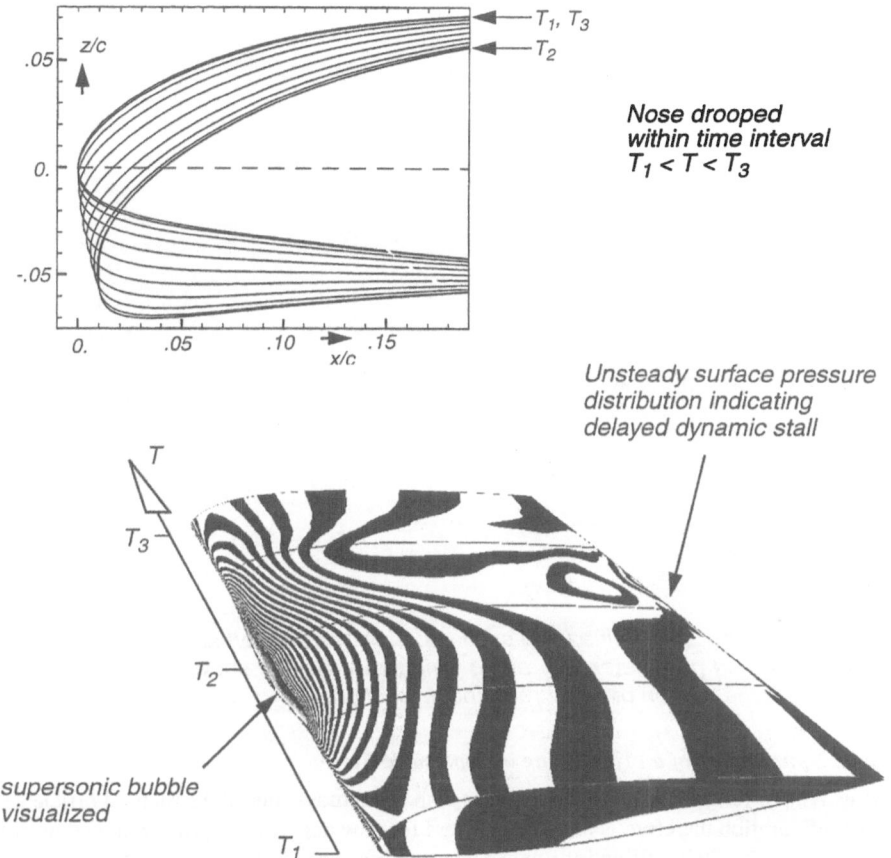

Nose drooped
within time interval
$T_1 < T < T_3$

Unsteady surface pressure
distribution indicating
delayed dynamic stall

supersonic bubble
visualized

*Fig. 9: Results for unsteady airfoil flow with periodically drooped nose (Ref. 15): Color isofringes for surface pressure visualization, $M_\infty = 0.3$. Control of locally occurring supersonic flow and viscous interaction triggering downstream boundary layer separation.*

*Airfoil parameters for wing geometry definition*

In a new modification of our wing generator software, the previously used support airfoils at selected stations along wing span may now be replaced by defining a set of airfoil parameters like those explained above for the PARSEC family, as functions along span, just like the already operational way of defining functions for leading and trailing edge coordinates and local wing section twist. Compared to the amount and flexibility of input data for a set of special support sections, the new approach seems to open a better use especially for optimization strategies, using the well proven concept of composing arbitrary spanwise curves for geometry and distributions with only a few key parameters. Figure 10 illustrates both options to define wing sections at any spanwise station in the wing coordinate system $(x, y_o)$.

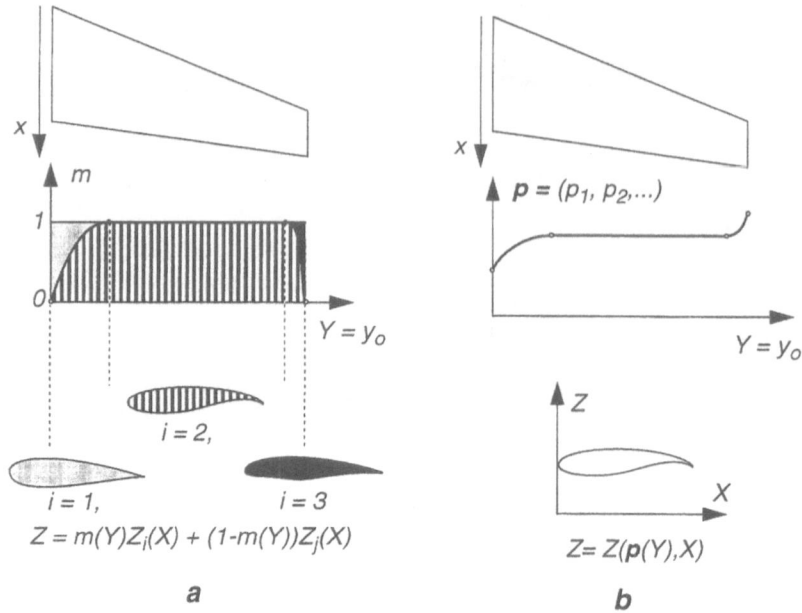

$$Z = m(Y)Z_i(X) + (1-m(Y))Z_j(X)$$

$$Z = Z(p(Y),X)$$

**a**       **b**

*Fig. 10: Two methods to define wing sections: Blending support airfoils data (a), and varying generating parameters (b) along wing span. Sketch shows wing with basic section over large portion of wing, root and tip sections.*

### Example: Optimization of a Flying Wing in Supersonic Flow

A Flying Wing aircraft has several advantages both for aerodynamic and structural efficiency. Such a configuration therefore is an ideal test case for new design and optimization strategies. The application of a Flying Wing for a new concept of high speed transport aircraft recently has been studied as an alternative to conventional supersonic configurations [16]. Using variable sweep angles for the whole configuration results in a wide range of possible optimum aerodynamic efficiency. Oblique Flying Wing (OFW) examples have been generated by our geometry preprocessor, both using the previous support section blending technique and also the new parametric 'PARSEC' airfoils defined along span of the OFW.

Several aerodynamic phenomena suggest usage of well-known design methods like applying transonic (supercritical) airfoil theory, applied to a sufficiently thick 2D basic airfoil in the subsonic Mach number component normal to the leading edge of this large aspect ratio wing. Reqirements of an elliptic load distribution suggests the classic elliptic planform, a linear variation of the wing twist along span and blending the basic airfoil with two modified sections at both tips yields a wing with the desired load distribution. With computed lift over drag ratio (L/D) of this first case study [17]slightly above the values of known conventional wing-body type configurations, we learned that a more refined approach than designing one 2D basic section would allow for a better control of crossflow shocks coalescing on the upper wing surface. With a slight modification of the planform geometry and the new spanwise parametric section defnition this goal has been accomplished already in a first manual approach of optimizing the OFW [19].

i

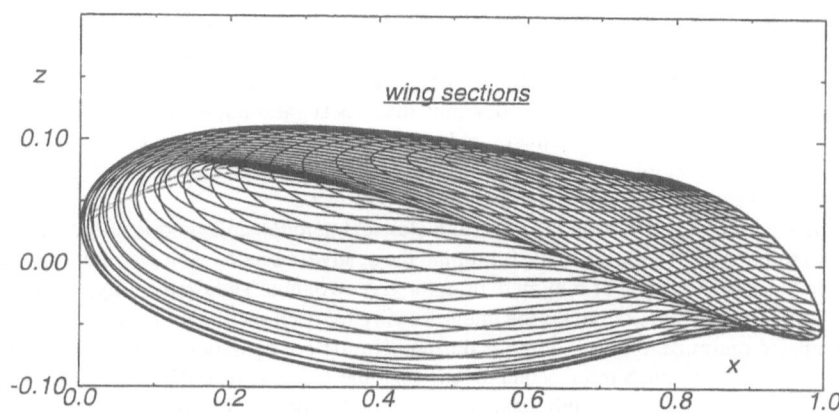

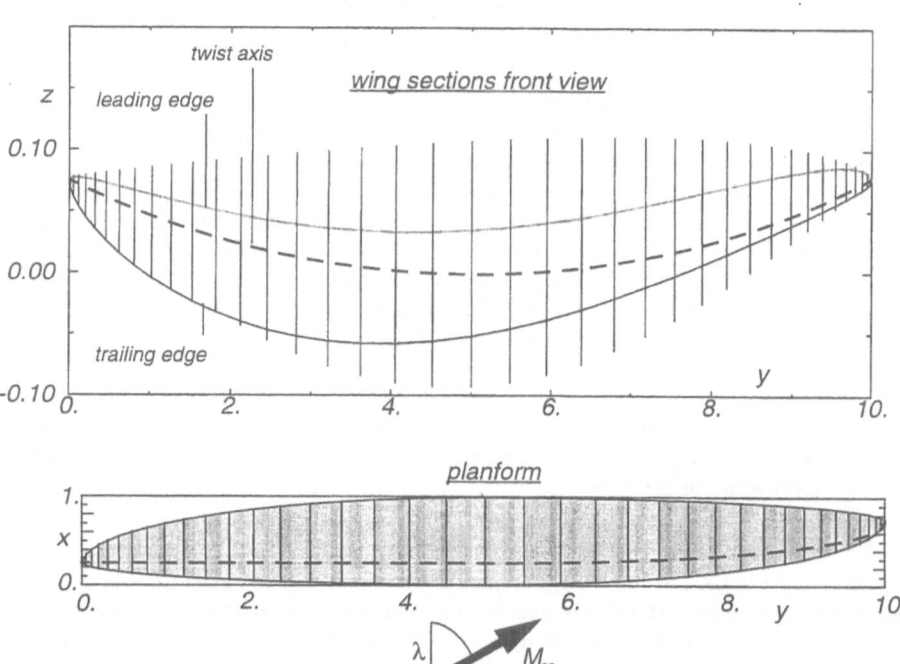

Fig. 11: Oblique Flying Wing optimized for supersonic flow $M_\infty = 1.4$: Example for spanwise variation of wing sections (above), dihedral, thickness and twist, leading and trailing edge geometry (center and below). (Note enlarged scale for vertical co-ordinate z). Aerodynamic performance optimized in swept flow $M_\infty = 1.4$, $\lambda = 60°$, with constraints for spanwise wing section thickness, area and aerodynamic load distribution.

Figure 11 shows the resulting geometry in a threeview: wing sections vary considerably within the elliptic chord distribution, all 11 PARSEC parameters were made functions of span. A modification of the basic elliptic planform to the unsymmetrical shape with stronger sweep at the trailing tip ($y = 10$) than at the leading tip ($y = 0$) is suggested by the observed stronger crossflow shocks in the trailing area. Simple aerodynamic theories suggest higher normal Mach number components in the trailing part and therefore lead us to shape the local sections for supercritical flow in higher Mach numbers. This was a goal in the example Fig. 3, a use of PARSEC functions for spanwise section definition therefore was promising.

Constraints based on application of one of the classical aerodynamic theories (the supersonic area rule) to improve the design, calls for tuning the spanwise section area distribution according to the Sears-Haack body of minimum drag for given volume. The simple polynomial structure of the PARSEC function yields the integral easily for each set of parameters.

An automated optimization procedure for this and similar configurations will perform the design of a better OFW much more economically than the manual approach done so far, but the value of learning the role of the individual parameters in the process of a practical design cannot be estimated high enough.

## Wings with multicomponent high lift system

In the same way as for airfoil parameters, the newly developed key points for sealed flaps and slats as well as the input data for multicomponent airfoil shapes may be made functions along the third physical dimension in space, the spanwise direction of a wing. 3D space is defined in the 'wing system', with planform in an $(x, y_o)$-plane, and the wing shape defined prior to shifting and rotation in general 3D space with aircraft coordinates.

Using airfoil and high lift sections data in the retracted (cruise conditions) position defining clean wing boundary conditions, wing input data rescale each section plus its components to physical chord and provides twist as a function of span. Each component (flap, slat, solid remaining wing) will then be available for a movement in 3D space.

These surfaces may be defined along most of the wing span (except at fat root fillets or at thin tips), choice of sections to begin and end slats and flaps is then a matter of constraints and additional flexibility. Kinematic requirements, however, in the case of tapered and twisted wings demand that the sectioning between component ends must be redefined to allow an unobstructed sliding of the flaps and slats along the fixed part of the wing. In the general case this requires an intersection of the wing surface with a sphere, its center located at the vertex of a cone tangent to the local wing surface panels.

### Example: An extended DLR-F5 test wing configuration

A decade ago the 'DLR-F5 wing' was presented and communicated as a test case for the development of CFD methods [2], [3]. The data for this wing have been used by various developers to tune Navier-Stokes codes for viscous transonic flow. The case is still a difficult task to solve if the experiment is to be simulated: transonic flow with laminar shock - boundary layer interaction remains a problem for CFD so far.

Nevertheless, with the example well known in the CFD community it seems to be a suitable example also for other than experimental operating conditions, especially if the wing shape is defined in a parametric way allowing variation of the shape and this way testing design and optimization strategies.

In a first approach to revisit the DLR-F5 case its wing sections are redefined by PARSEC parameters. The original wing has a symmetrical basic section which was designed to be nearly shock-free at $M_\infty = 0.78$. A set of surface data points was provided and blended with a thick NACA 0036 section to form a prominent wing root fillet. With knowing the nose and crest curvatures, applying the trailing edge modification parameters and blending with the NACA 4digit generating function, the input set of airfoil data can be replaced altogether by the new analytical definition. The new parameters are proposed along with mathematical modelling of some experimental pressure distributions to complete a new test case for direct/inverse CFD and for optimization [19].

In addition, the DLR-F5 wing was used for definition of a multicomponent wing with slat and flap. Figure 12 (a) shows the choice for carving the basic section to shape a slat and a flap, these section components subsequently are scaled to the DLR-F5 planform and additional input for the 3D flap and slat tracks and rotation angles is provided. Choice of spanwise extent for flap and slat, a refined sliding sections definition and closure of the components at the sliding sections completes the preprocessing of this extended test wing for CFD analysis.

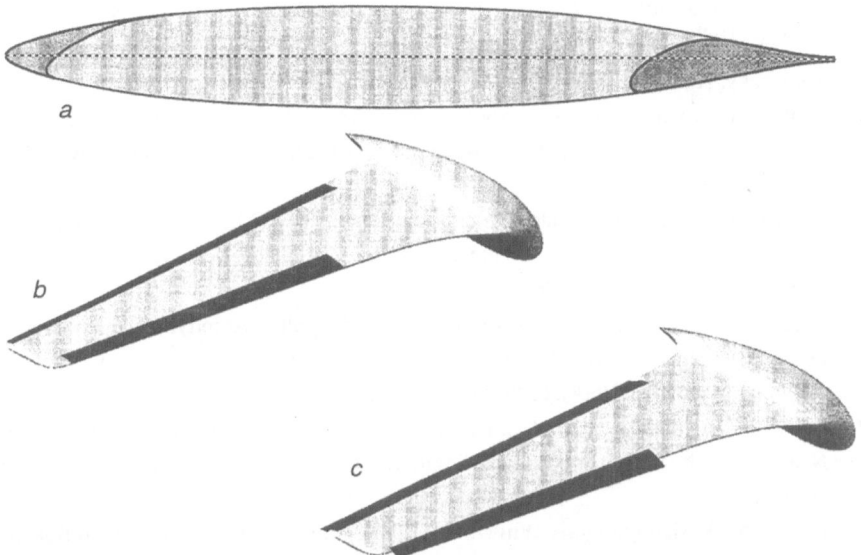

Fig. 12: Basic DLR-F5 section, redefined by PARSEC parameters and carved to include a slat and flap component (a). Selection of 3D flap and slat extending along span, in clean wing (b) and high lift (c) position.

## Conclusions

An effort is made in using basic algebraic and analytic relations to generate realistic airfoil shapes which are specified from a set of parameters. These are defined by only a few characteristic dimensions used already in classical airfoil catalogs like NACA airfoil families, but also allow for a refined shape definition as it results from systematic design processes in the transonic flight regime. Airfoils determined this way by a minimum set of parameters are subsequently used as wing sections, with their generating parameters made functions along span, This has been proven earlier for basic wing geometries which have used given data sets for support airfoils. Some unsteady airfoil flow applications lead the way to fully threedimensional wings which may be subject to input for manual or automated aerodynamic optimization techniques. This method to describe all shapes analytically has been extended to high lift systems and adaptive devices.

The approach is intended to provide 2D, 3D and, with unsteady, adaptive or evolutionary configurations, also 4D boundary conditions for CFD and CAD.

# References

[1] Sobieczky, H.: Computational Methods for the Design of Adaptive Airfoils and Wings. Notes on Numerical Fluid Mechanics, Vol. 2, ed. E. H. Hirschel, Vieweg (1979) pp. 269 - 278.

[2] Sobieczky, H.: DFVLR-F5 Test Wing Configuration for Computational and Experimenta Experimental Aerodynamics, Wing Surface Generator Code, Control Surface and Boundary Conditions. Notes on Num. Fluid Mechanics, Vol. 22, ed. W. Kordulla, Vieweg (1988), pp. 27 - 37.

[3] Sobieczky, H., Hefer, G., Tusche, S.: DFVLR-F5 Test Wing Experiment for Computational Aerodynamics. AIAA 5. Appl. Aerodynamics Conf. Proc., AIAA 87-2485CP, (1987), Notes on Numerical Fluid Mechanics, Vol. 22, ed. W. Kordulla, Vieweg, (1988), pp.4 - 22.

[4] Sobieczky, H.: Aircraft Surface Generation. Results of EC Brite/Euram Project 'Euromesh' 1990-92, Notes on Numerical Fluid Mechanics, Vol. 44, ed. N. Weatherill et al, Vieweg (1993), pp. 71 - 76.

[5] Sobieczky, H.: Geometry Generator for CFD and Applied Aerodynamics. in: New Design Concepts for High Speed Air Transport. CISM Courses and Lectures No. 366, Springer (Wien, NewYork), (1997) pp 137 - 157.

[6] Ladson, C., Brooks, C.: Development of a Computer Program to obtain Ordinates for NACA 4-Digit, 4-Digit Modified, 5-Digit, and 16-Series Airfoils. NASA TM X-3284 (1975).

[7] Eberle, A.: Berichtigung der Whitcomb-Offenlegungsschrift und Profildefinitionsprogramm. MBB-UFE-AERO-MT-298 (1977).

[8] Zores, R.: Transonic Airfoil Design with Expert Systems. AIAA 95-1818CP, Proc. 13th AIAA Applied Aerodynamics Conf., San Diego, CA, (1995).

[9] Garabedian, P. R.: On the Design of Airfoils Having no Boundary Layer Separation. Advances in Mathematics 15, (1975) pp 164-168.

[10] Henne, P. A.: Innovation with Computational Aerodynamics: The Divergent Trailing Edge Airfoil", Applied Computational Aerodynamics, ed. P. A. Henne, AIAA Education Series, AIAA, Washington, D.C., (1990), pp. 221 - 262.

[11] Thompson, B. E., Lotz, R. D.: Divergent-Trailing-Edge Airfoil Flow. J. Aircraft Vol. 33 (1996), pp. 950 - 955.

[12] Sobieczky, H.: Gasdynamic Knowledge Base for High Speed Flow Modelling. in: New Design Concepts for High Speed Air Transport. CISM Courses and Lectures No. 366,

Springer (Wien, NewYork), (1997) pp 105 - 119.

[13]  Ashill, P. R., Fulker, J. L., Simmons, M. J., Gaudet, I. M.: A Review of Research at DRA on Active and Passive Control of Shock Waves. 20th ICAS Congress Conf. Proc. ICAS-96-2.1.4, (1996) pp. 76 - 87.

[14]  Geissler, W., Sobieczky, H.: Unsteady Flow Control on Rotor Airfoils. AIAA 95-1890CP, Proc. 13th AIAA Applied Aerodynamics Conference, San Diego, CA, (1995).

[15]  Sobieczky, H., Geissler, W., Hannemann, M.: Numerical Tools for Unsteady Viscous Flow Control". Proc. 15th Int. Conf. on Num. Meth. in Fluid Dynamics. Lecture Notes in Physics, ed. P. Kutler, J. Flores, J.-J. Chattot, Springer (Berlin, Heidelberg) (1997).

[16]  Van der Velden, A.: The Oblique Flying Wing Transport. in: New Design Concepts for High Speed Air Transport. CISM Courses and Lectures No. 366, Springer (Wien, NewYork), (1997) pp 291 - 316.

[17]  Seebass, R.: Oblique Flying Wing Studies. in: New Design Concepts for High Speed Air Transport. CISM Courses and Lectures No. 366, Springer (Wien, NewYork), (1997) pp 317 - 336.

[18]  Li, P., Seebass, R., Sobieczky, H.: Manual Optimization of an Oblique Flying Wing. AIAA 98-0598 (1998).

[19]  Sobieczky, H.: Configuration Test Cases for Aircraft Wing Root Design and Optimization. Proc. ISIP'98, Nagano, Japan, (1998).

# Using Existing Flow-Field Analysis Codes for
# Inverse Design of Three-Dimentional Aerodynamic Shapes

George S. Dulikravich and Daniel P. Baker

Department of Aerospace Engineering
The Pennsylvania State University, University Park, PA 16802, U.S.A.
e-mail: ft7@psu.edu

## Summary

This chapter demonstrates that it is possible to use existing proven arbitrary flow-field analysis codes without any modifications to perform inverse aerodynamic design of two-dimensional and three-dimensional shapes. A general concept for inverse design of aerodynamic shapes based on treating the aerodynamic surface as an elastic membrane subject to a specified surface pressure distribution is explained. A new mathematical formulation based on the Fourier series analytical solution of this model is detailed. The method is formulated for two-dimensional and for three-dimensional configurations. It can be used in conjunction with any available flow-field analysis code without a need for modification of such a code. Preliminary testing of the new method is performed with a potential flow surface panel code, an Euler flow solver, and a Navier-Stokes flow solver. The convergence rate of the design process is found to be similar for both non-lifting and lifting aerodynamic shapes with the Navier-Stokes flow solver typically requiring minimum number of design iterations. Suggestions for further research and improvements are made.

## Introduction

Aerodynamic shape inverse design methods have the goal of determining the proper shape of aerodynamic body such that a desired (target) pressure distribution is achieved on its surface. There are several methods [1,2,3] that are capable of such inverse determination of domain size and shape, but most of the methods require the development of new complex mathematical formulations and the accompanying new software. The majority of inverse design methods require at least the modification of boundary conditions enforcement subroutines. This is often non-trivial and even impossible to do if the source code is not available to the designer. Therefore, inverse shape design methods that require a minimum of software development and can accept any existing reliable flow-field analysis computer code as a large interchangeable subroutine are highly desirable.

This chapter focuses on one such formulation. The so-called elastic surface membrane concept was first proposed by Garabedian and McFadden [4,5] who considered the surface of an aerodynamic body to deform under aerodynamic loads in a manner similar to an elastic membrane. Their method was then adapted by Malone et al. [6-9] into what is presently known as the MGM (modified Garabedian-McFadden or Malone-Garabedian-McFadden) technique.

The idea behind this method is to model the boundary of an aerodynamic body as a thin elastic membrane. The membrane is then subjected to a distributed normal forcing function represented by the local differences between target (specified) surface coefficients of pressure and those surface coefficients of pressure found by performing an aerodynamic flow-field analysis of the guessed body shape. Due to the membrane's elasticity, it then iteratively and smoothly deforms with time until reaching a steady state configuration, whereby the target pressure values are matched by those generated by the deformed shape. To model the damped, unsteady motion of the elastic surface, any artificial or non-physical governing equation can be used, relating local membrane deformations to the distribution of the forcing function on the surface of the membrane. One such model, as suggested by Garabedian and McFadden [4,5], is

$$\beta_0 \frac{\partial \Delta n}{\partial \tau} + \beta_1 \frac{\partial^2 \Delta n}{\partial x \partial \tau} + \beta_2 \frac{\partial^3 \Delta n}{\partial x^2 \partial \tau} = \Delta C_p \; . \tag{1}$$

Here, $\Delta n$ is the local outward normal deformation of the airfoil contour, $\tau$ is the time coordinate, x is the Cartesian spatial coordinate, while coefficients $\beta$ are user specified and of order one. The local difference between the target and the actual computed surface coefficients of pressure is

$$\Delta C_p = C_{p \; target} - C_{p \; actual} \; . \tag{2}$$

Since the objective is to determine the steady state condition of the aerodynamic forces, the time derivatives in Eq. (1) can be eliminated by discretizing the artificial time into equal steps corresponding to each iteration of the inverse design procedure. If these iterative time steps are treated as $\Delta \tau = 1$, then Eq. (1) becomes an ordinary differential equation with constant coefficients and a forcing function

$$\beta_0 \Delta n + \beta_1 \frac{d \Delta n}{dx} + \beta_2 \frac{d^2 \Delta n}{dx^2} = \Delta C_p \; . \tag{3}$$

A common modification to the shape evolution equation (3) is to allow deformations to occur only in the Cartesian y-direction, simplifying the update of geometries and allowing for faster computation. This approach, as used by Malone, assumes the governing equation

$$\beta_0 \Delta y + \beta_1 \frac{d \Delta y}{dx} + \beta_2 \frac{d^2 \Delta y}{dx^2} = \Delta C_p \; . \tag{4}$$

Traditionally, Eq. (4) is solved for the correction ($\Delta y$) in airfoil y-coordinates by discretizing the airfoil contour and utilizing finite differencing at each discretization point, i, on that contour to represent the first derivative and the second derivative in Eq. (4). After finite differencing at all of the surface membrane points, the result is a set of linear algebraic equations of the form

$$a_i \Delta y_{i-1} + b_i \Delta y_i + c_i \Delta y_{i+1} = \Delta C_{p_i} \; . \tag{5}$$

In the general case when the contour discretization points are unevenly spaced, the finite difference formulas applied at any point on the airfoil upper contour result in

$$a_i = \frac{\beta_1}{x_i - x_{i-1}} - \frac{2\beta_2}{(x_i - x_{i-1})(x_{i+1} - x_{i-1})} \qquad (6)$$

$$b_i = -\beta_0 - \frac{\beta_1}{x_k - x_{k-1}} - \frac{\beta_2}{(x_k - x_{k-1})(x_{k+1} - x_k)} \qquad (7)$$

$$c_i = -\frac{2\beta_2}{(x_{i+1} - x_{i-1})(x_{i+1} - x_i)} \qquad (8)$$

while on the airfoil lower contour the result is

$$a_i = \frac{2\beta_2}{(x_{i+1} - x_{i-1})(x_i - x_{i-1})} \qquad (9)$$

$$b_i = \beta_0 - \frac{\beta_1}{x_{i+1} - x_i} - \frac{\beta_2}{(x_i - x_{i-1})(x_{i+1} - x_i)} \qquad (10)$$

$$c_i = \frac{\beta_1}{x_{i+1} - x_i} - \frac{2\beta_2}{(x_{i+1} - x_i)(x_{i+1} - x_{i-1})} . \qquad (11)$$

The tri-diagonal system (Eq. 5) can easily and efficiently be solved using the Thomas algorithm. To avoid the ambiguity of the upper and lower contour finite difference equations, one approach is to fix the trailing edge point.

$$\Delta y_{TE} = 0 \qquad (12)$$

and to make the motion of the leading edge node as the average of the displacements of the two nodes adjacent to it

$$\Delta y_{\text{leading edge}} = \frac{\Delta y_{\text{leading edge-1}} + \Delta y_{\text{leading edge+1}}}{2} . \qquad (13)$$

One major problem with the classical MGM approach is its slow convergence at the leading and trailing edges of the airfoil, as compared to the mid-chord regions of the airfoil. Another major problem is the governing equation's non-physical, *ad hoc* nature. Furthermore, there is no analytical method to determine the optimum coefficients $\beta_0$, $\beta_1$, $\beta_2$ in Eq. (4), although their choice can radically change the convergence of the MGM inverse shape design process.

In order to improve the quality of the solution of Eq. (4), higher order accurate finite difference schemes can be implemented by increasing the stencil size used in the calculation of the derivatives of y-coordinate corrections. This is most easily done by rediscretizing the airfoil surface such that equidistant points along the airfoil surface contour are used. Unfortunately, all benefits of grid clustering are lost in the process. Because the stencil size

91

for all derivatives is increased with improved order of accuracy, stencils for derivatives at points near the leading and trailing edges of the airfoil can then be made to wrap around to the other side of the airfoil. This also aids to the smoothness of leading and trailing edge deformation. In general, the more neighboring points used in the determination of derivatives at each grid index, the more accurate the derivative. However, the effects of increasing the derivative order of accuracy are negligible or even detrimental once a certain order of accuracy has been reached. From numerical experimentation, the optimum stencil size in most cases was found to be five. When more than five points are used in the determination of derivatives in Eq. (4), all benefits of increased order of accuracy are opposed by errors caused by numerical singularity in the coefficient matrix. A disadvantage of larger stencil size is that the coefficient matrix multiplying the nodal displacements is no longer tri-diagonal, and must be inverted in a more computationally intensive manner. Using a singular value decomposition algorithm [10] to invert the matrix and minimize the effect of singularities (which occur with higher frequency as the derivative order of accuracy is increased) requires a number of computations proportional to the cube of the number of surface nodes. In comparison, tri-diagonal systems of equations can be solved with calculations on the order of the number of equations to the first power. Thus, the higher order accurate process is significantly slower per iteration than the traditional MGM approach.

In an attempt to counter these problems while improving the convergence rate of the design process, a new method of solution of the elastic membrane equation has been devised [11,12]. It is based on the transformation of the derivatives with respect to x-coordinate in Eq. (4) to derivatives with respect to airfoil contour-following coordinate, s. This method uses an analytical solution of the shape evolution equation in terms of a Fourier series.

## Fourier Series Solution of the Two-dimensional Shape Evolution Equation

The analytical solution of the MGM equation given in Eq. (4) is complicated by the fact that the $\beta_0$ and $\beta_2$ terms switch signs when moving from the bottom surface of the airfoil to the top surface (or vice versa). If $\beta_0, \beta_1, \beta_2$ are considered to be positive constants, Eq. (4) takes the following forms on the top and bottom surfaces of an airfoil:

Top Surface: $\qquad \beta_0 \Delta y + \beta_1 \dfrac{d\Delta y}{ds} - \beta_2 \dfrac{d^2 \Delta y}{ds^2} = \Delta C_p(s)$ $\qquad\qquad\qquad$ (14)

Bottom Surface: $\quad -\beta_0 \Delta y + \beta_1 \dfrac{d\Delta y}{ds} + \beta_2 \dfrac{d^2 \Delta y}{ds^2} = \Delta C_p(s)$ . $\qquad\qquad$ (15)

Both of these equations can be considered as the generalized mass-damper-spring equation.

$\qquad k\Delta y + c \dfrac{\partial \Delta y}{\partial s} + m \dfrac{\partial^2 \Delta y}{\partial s^2} = \Delta C_p$ . $\qquad\qquad\qquad\qquad\qquad$ (16)

Here, the time coordinate has been replaced with the surface following coordinate, s, and the forcing function $\Delta C_p(s)$ is an arbitrary function of the coordinate s. The homogeneous solution of Eq. (16) can be found by assuming

92

$$\Delta y_h = e^{\lambda s}. \tag{17}$$

On the bottom surface of the airfoil

$$k^{\text{bottom}} = \beta_0 \tag{18}$$

$$c = \beta_1 \tag{19}$$

$$m^{\text{bottom}} = -\beta_2 \cdot \tag{20}$$

This leads to

$$\lambda_{1,2}^{\text{bottom}} = \frac{-\beta_1 \pm \sqrt{\beta_1^2 + 4\beta_0\beta_2}}{-2\beta_2} \tag{21}$$

$$\Delta y_h^{\text{bottom}} = F^{\text{bottom}} e^{\lambda_1^{\text{bottom}} s} + G^{\text{bottom}} e^{\lambda_2^{\text{bottom}} s}. \tag{22}$$

On the top surface of the airfoil, the signs of k and m reverse

$$k^{\text{top}} = -\beta_0 \tag{23}$$

$$m^{\text{top}} = \beta_2 \cdot \tag{24}$$

This leads to

$$\lambda_{1,2}^{\text{top}} = \frac{-\beta_1 \pm \sqrt{\beta_1^2 + 4\beta_0\beta_2}}{2\beta_2} \tag{25}$$

$$\Delta y_h^{\text{top}} = F^{\text{top}} e^{\lambda_1^{\text{top}} s} + G^{\text{top}} e^{\lambda_2^{\text{top}} s} \tag{26}$$

where F and G are (as yet) undetermined coefficients. The particular solution of Eq. (16) can be found by creating a Fourier series expansion of the function $\Delta C_p(s)$

$$\Delta C_p(s) = a_0 + \sum_n \left[ a_n \cos N_n s + b_n \sin N_n s \right] \tag{27}$$

where

$$N_n = \frac{n\pi}{L} \cdot \tag{28}$$

Here, L is one-half of the total arc length of the airfoil contour. A particular solution is assumed of the form

$$\Delta y_p = A_0 + \sum_{n=1}^{\infty} [A_n \cos N_n s + B_n \sin N_n s] \tag{29}$$

$$\frac{\partial \Delta y_p}{\partial s} = \sum_{n=1}^{\infty} [-A_n N_n \cos N_n s + B_n N_n \sin N_n s] \tag{30}$$

$$\frac{\partial^2 \Delta y_p}{\partial s^2} = -\sum_{n=1}^{\infty} [A_n N_n^2 \cos N_n s + B_n N_n^2 \sin N_n s] \ . \tag{31}$$

Substitution of Eq. (27) and Eqs. (29-31) into the general evolution equation (16) and collection of like terms yields

$$A_0 = \frac{a_0}{k} \tag{32}$$

$$A_n = \frac{a_n(k - N_n^2 m) - b_n(cN_n)}{(k - N_n^2 m)^2 + (cN_n)^2}, n = 1,2,3,\cdots \tag{33}$$

$$B_n = \frac{b_n(k - N_n^2 m) + a_n(cN_n)}{(k - N_n^2 m)^2 + (cN_n)^2}, n = 1,2,3,\cdots \ . \tag{34}$$

Thus, the complete solution for $\Delta y$ on the top and bottom surface of the airfoil is

$$\Delta y = F e^{\lambda_1 s} + G e^{\lambda_2 s} + A_0 + \sum_{n=1}^{\infty} [A_n \cos N_n s + B_n \sin N_n s] \ . \tag{35}$$

The unknown constants, F and G, on the top and bottom surfaces are determined by specifying four boundary conditions. The following four conditions can be used: trailing edge closure, leading edge closure, zero trailing edge displacement, and smoothness of $\Delta y$ at the leading edge. For trailing edge closure condition can be expressed as

$$\Delta y^{\text{bottom}}(0) = \Delta y^{\text{top}}(2L) \ . \tag{36}$$

For pinned trailing edge,

$$\Delta y^{\text{bottom}}(0) = 0 \ . \tag{37}$$

The combination of Eq. (35) and Eq. (36) yields the following boundary condition equation

$$F^{\text{bottom}} + G^{\text{bottom}} = -\sum_{n=0}^{\infty} A_n^{\text{bottom}} \ . \tag{38}$$

Similarly, the combination of Eq. (35) and Eq. (37) yields the following boundary condition equation

$$F^{top}e^{2L\lambda_1^{top}} + G^{top}e^{2L\lambda_2^{top}} = -\sum_{n=0}^{\infty}A_n^{top}. \tag{39}$$

The leading edge closure condition

$$\Delta y^{bottom}(s_{LE}) = \Delta y^{top}(s_{LE}) \tag{40}$$

can be expressed as

$$F^{bottom}e^{s_{LE}\lambda_1^{bottom}} + G^{bottom}e^{s_{LE}\lambda_2^{bottom}} - F^{top}e^{s_{LE}\lambda_1^{top}} - G^{top}e^{s_{LE}\lambda_2^{top}}$$
$$= \Delta y_p^{top}(s_{LE}) - \Delta y_p^{bottom}(s_{LE}). \tag{41}$$

The smooth leading edge deformation condition

$$\frac{d}{ds}\Delta y^{bottom}(s_{LE}) = \frac{d}{ds}\Delta y^{top}(s_{LE}) \tag{42}$$

can be expressed as

$$F^{bottom}\lambda_1^{bottom}e^{s_{LE}\lambda_1^{bottom}} + \lambda_2^{bottom}G^{bottom}e^{s_{LE}\lambda_2^{bottom}}$$
$$- F^{top}\lambda_1^{top}e^{s_{LE}\lambda_1^{top}} - G^{top}\lambda_2^{bottom}e^{s_{LE}\lambda_2^{top}}$$
$$= \frac{d}{ds}\Delta y_p^{top}(s_{LE}) - \frac{d}{ds}\Delta y_p^{bottom}(s_{LE}). \tag{43}$$

F and G coefficients can be found by simultaneous solution of Eqs. (38, 39, 41, 43).

$$\begin{Bmatrix} F^{bottom} \\ G^{bottom} \\ F^{top} \\ G^{top} \end{Bmatrix} = \begin{bmatrix} 1 & 1 & 0 & 0 \\ 0 & 0 & e^{2L\lambda_1^{top}} & e^{2L\lambda_2^{top}} \\ e^{s_{LE}\lambda_1^{bottom}} & e^{s_{LE}\lambda_2^{bottom}} & -e^{s_{LE}\lambda_1^{top}} & -e^{s_{LE}\lambda_2^{top}} \\ \lambda_1^{bottom}e^{s_{LE}\lambda_1^{bottom}} & \lambda_2^{bottom}e^{s_{LE}\lambda_2^{bottom}} & -\lambda_1^{top}e^{s_{LE}\lambda_1^{top}} & -\lambda_2^{top}e^{s_{LE}\lambda_2^{top}} \end{bmatrix}^{-1} \cdot$$

$$\begin{Bmatrix} -\sum_{n=0}^{\infty}A_n^{bottom} \\ -\sum_{n=0}^{\infty}A_n^{top} \\ \Delta y_p^{top}(s_{LE}) - \Delta y_p^{bottom}(s_{LE}) \\ \Delta y_p^{top\prime}(s_{LE}) - \Delta y_p^{bottom\prime}(s_{LE}) \end{Bmatrix}. \tag{44}$$

Since the Fourier series formulation is exact, any errors due to finite differencing and the need for Eq. (13) are removed. The choice of number of Fourier series terms effectively enforces an upper limit on the frequency allowed in the y-coordinate deformation.

## Solving the Three-dimensional Shape Evolution Equation

Generalization of the elastic membrane formulation to surfaces of three-dimensional objects can be accomplished by a complete form of a second order partial differential equation of the following type [12]

$$a\frac{\partial^2 \Delta y}{\partial t^2} + b\frac{\partial^2 \Delta y}{\partial s \partial t} + c\frac{\partial^2 \Delta y}{\partial s^2} + d\frac{\partial \Delta y}{\partial s} + e\frac{\partial \Delta y}{\partial t} - f\Delta y = F(\Delta C_p). \tag{45}$$

Here, s and t are the surface coordinates in the general streamwise and spanwise directions, respectively. The forcing function can be generalized as

$$F(\Delta C_p) = \Delta C_p + \alpha_1\frac{\partial \Delta C_p}{\partial s} + \beta_1\frac{\partial \Delta C_p}{\partial t} + \alpha_2\frac{\partial^2 \Delta C_p}{\partial s^2} + \beta_2\frac{\partial^2 \Delta C_p}{\partial t^2}. \tag{46}$$

Here, the coefficients a, b, c, d, e, f, $\alpha$ and $\beta$ are user specified. The coefficients have been found [4-9, 11,12] to influence the convergence rate of the iterative shape design process. The general variation of the surface forcing function, F, can be represented as a Fourier series

$$F(C_p) = \sum_{n=0}^{n_{max}} \sum_{m=0}^{m_{max}} [A_{mn}\cos(ns)\cos(mt) + B_{mn}\cos(ns)\sin(mt)$$

$$+ C_{mn}\sin(ns)\cos(mt) + D_{mn}\sin(ns)\sin(mt)]. \tag{47}$$

Similarly, the particular solution of the elastic membrane equation (45) can be represented in a Fourier series form as

$$\Delta y^P = \sum_{n=0}^{n_{max}} \sum_{m=0}^{m_{max}} [a_{mn}\cos(ns)\cos(mt) + b_{mn}\cos(ns)\sin(mt)$$

$$+ c_{mn}\sin(ns)\cos(mt) + d_{mn}\sin(ns)\sin(mt)]. \tag{48}$$

If the mixed second spatial derivative is kept in the general elastic membrane motion equation (45), the homogeneous part of the solution of this equation cannot be found directly. A linear transformation of coordinates [12] is needed in this case that somewhat complicates the process. For the sake of simplicity and the clarity of explanation, we will continue this elaboration with the simplified Eq. (45) and Eq. (46) by working with the assumption that

$$b = \alpha_1 = \beta_1 = \alpha_2 = \beta_2 = 0. \tag{49}$$

These terms can be omitted without any detrimental effects on the robustness and the accuracy of the entire design concept, but they may influence convergence rate of the iterative design process. Then, from Eq. (48) it follows that

$$\frac{\partial \Delta y^p}{\partial s} = \sum_{n=0}^{n_{max}} \sum_{m=0}^{m_{max}} -n[a_{mn} \sin(ns)\cos(mt) + b_{mn} \sin(ns)\sin(mt)$$

$$-c_{mn} \cos(ns)\cos(mt) - d_{mn} \cos(ns)\sin(mt)] \tag{50}$$

$$\frac{\partial^2 \Delta y^p}{\partial s^2} = \sum_{n=0}^{n_{max}} \sum_{m=0}^{m_{max}} -n^2[a_{mn} \cos(ns)\cos(mt) + b_{mn} \cos(ns)\sin(mt)$$

$$+c_{mn} \sin(ns)\cos(mt) + d_{mn} \sin(ns)\sin(mt)] \tag{51}$$

$$\frac{\partial \Delta y^p}{\partial t} = \sum_{n=0}^{n_{max}} \sum_{m=0}^{m_{max}} -m[a_{mn} \cos(ns)\sin(mt) - b_{mn} \cos(ns)\cos(mt)$$

$$+c_{mn} \sin(ns)\sin(mt) - d_{mn} \sin(ns)\cos(mt)] \tag{52}$$

$$\frac{\partial^2 \Delta y^p}{\partial t^2} = \sum_{n=0}^{n_{max}} \sum_{m=0}^{m_{max}} -m^2[a_{mn} \cos(ns)\cos(mt) + b_{mn} \cos(ns)\sin(mt)$$

$$+c_{mn} \sin(ns)\cos(mt) + d_{mn} \sin(ns)\sin(mt)] \ . \tag{53}$$

After substitution of Eqs. (47-53) into Eq. (45) and matching of coefficients multiplying each of the four products of Sine and Cosine functions, the coefficients in Eq. (45) can be expressed as analytical functions of the coefficients in Eq. (46). That is,

$$\begin{bmatrix} (-an^2 - cm^2 - f) & em & dn & 0 \\ -em & (-an^2 - cm^2 - f) & 0 & dn \\ -dn & 0 & (-an^2 - cm^2 - f) & -em \\ 0 & dn & -em & (-an^2 - cm^2 - f) \end{bmatrix} \begin{Bmatrix} a_{mn} \\ b_{mn} \\ c_{mn} \\ d_{mn} \end{Bmatrix}$$

$$= \begin{Bmatrix} A_{mn} \\ B_{mn} \\ C_{mn} \\ D_{mn} \end{Bmatrix} \ . \tag{54}$$

Thus, the particular solution of Eq. (45) can be found in terms of the coefficients $A_{mn}, B_{mn}, C_{mn}, D_{mn}$ defining the pressure forcing function distribution on the surface of the elastic membrane.

If, in addition to the simplifications listed in Eq. (49), we further simplify the model by using

$$d = e = 0 \tag{55}$$

then the elastic membrane equation reduces to a very simple model given by Eq. (56)

$$a\frac{\partial^2 \Delta y}{\partial s^2} + c\frac{\partial^2 \Delta y}{\partial t^2} - f\Delta y = \Delta C_p \ .$$

(56)

The particular solution of Eq. (56) can be expressed as

$$\Delta y^P = \sum_{n=0}^{n_{max}} \sum_{m=0}^{m_{max}} \{[-A_{mn}\cos(ns)\cos(mt) - B_{mn}\cos(ns)\sin(mt)$$

(57)

$$- C_{mn}\sin(ns)\cos(mt) - D_{mn}\sin(ns)\sin(mt)]/(an^2 + cm^2 + f)\} \ .$$

This simple expression is very easy to program. It is capable of generating three-dimensional wing shapes involving both aerodynamic twist and geometric twist variations. When the difference between the specified and the calculated surface pressure has abrupt variations as in the case of a shock wave, this formulation creates a surface depression underneath the shock that oscillates along the wing surface as an undamped surface wave. Consequently, for design of shocked transonic configurations, the simplifications listed in Eq. (54) should not be used.

## Inverse Design of a Rectangular Patch of the Surface

If it is desirable to modify only one portion of an aerodynamic surface while maintaining the original shape of the rest of the body, the following formulation should be applied.

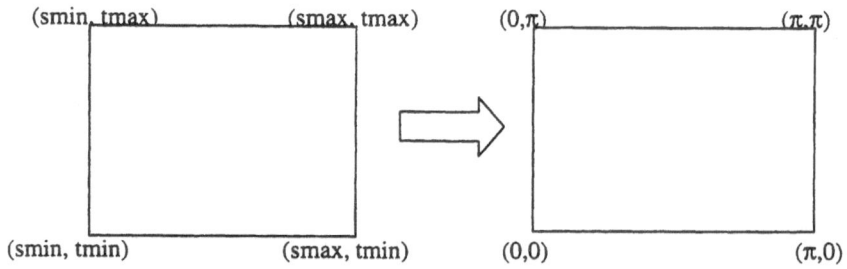

Fig. 1: Coordinate Mapping for Design of a Quadrilateral Surface Patch.

In order to maintain continuity of the surface, $\Delta y$ at each patch boundary must be fixed at zero. This can be accomplished by scaling the patch s and t coordinates to a rectangle with coordinates as shown in Fig 1, and using a Fourier Sine series (that is, $a_{mn}$, $b_{mn}$, $c_{mn}$ =0) to represent the deformation function $\Delta y_p(s,t)$. Then, the particular solution automatically satisfies all boundary conditions, and there is no need for a homogeneous solution.

$$\Delta y(s,t) = \sum_{n=1}^{n_{max}} \sum_{m=1}^{m_{max}} \frac{-D_{mn}\sin(ns)\sin(mt)}{an^2 + cm^2 + f} \ .$$

(58)

## Inverse Design of Three-dimensional Wings

For wing shape design it is desirable to enforce the following boundary conditions on the displacement $\Delta y$:
1. The displacement should be symmetric about the root section (vertical symmetry plane).
2. The displacement should be periodic in the s-direction for each span station.
3. The span-wise derivative of the displacement should be zero at the root section.
4. One point should be fixed. For example, the trailing edge of the root section can be fixed.
Firstly, scaling of the s and t coordinates of the wing should be performed as shown in Fig. 2.

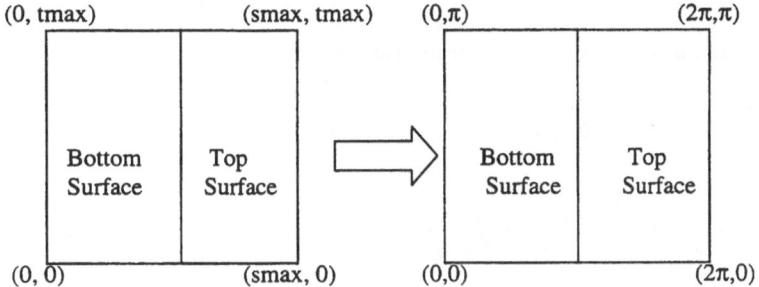

**Fig. 2: Coordinate Mapping for Design of a Three-dimensional Wing.**

Then, the four above boundary conditions can be mathematically expressed as:

$$\Delta y(s,t) = \Delta y(s,-t) \tag{59}$$

$$\Delta y(0,t) = \Delta y(2\pi n, t) \tag{60}$$

$$\Delta y_t(s,0) = 0 \tag{61}$$

$$\Delta y(0,0) = 0 . \tag{62}$$

A Cosine series in the t-direction with Sine and Cosine terms in the s-direction for the particular solution with assumptions listed in Eq. (49) and Eq. (55) automatically satisfies the first three boundary conditions.

$$\Delta y^P(s,t) = \sum_{n=0}^{n_{max}} \sum_{m=0}^{m_{max}} \frac{-A_{mn} \cos(ns)\cos(mt) - C_{mn}\sin(ns)\cos(mt)}{an^2 + cm^2 + f} . \tag{63}$$

Using a solution of the homogeneous part of the simplified elastic membrane equation can satisfy the remaining boundary conditionIf $\Delta y^h$ is a function of coordinate t only, the solution can be expressed as

$$\Delta y^h = F\cosh t\sqrt{\frac{f}{a}} + G\sinh t\sqrt{\frac{f}{a}} . \tag{64}$$

For zero displacement at the wing root trailing edge point,

$$\Delta y^h(0) = -\Delta y^p(0,0) = \sum_{n=0}^{n_{max}} \sum_{m=0}^{m_{max}} \frac{A_{mn}}{an^2 + cm^2 + f} \tag{65}$$

$$\Delta y^h(0) = F \tag{66}$$

$$F = \sum_{n=0}^{n_{max}} \sum_{m=0}^{m_{max}} \frac{A_{mn}}{an^2 + cm^2 + f} \ . \tag{67}$$

For zero slope of displacement in the t-direction at the root,

$$\frac{\partial \Delta y^h}{\partial t}(0) = -\frac{\partial \Delta y^p}{\partial t}(0,0) = 0 \tag{68}$$

$$\frac{\partial \Delta y^h}{\partial t} = (F \sinh t \sqrt{\frac{f}{a}} + G \cosh t \sqrt{\frac{f}{a}}) \sqrt{\frac{f}{a}} \tag{69}$$

$$\frac{\partial \Delta y^h}{\partial t}(0) = G \sqrt{\frac{f}{a}} \tag{70}$$

$$G = 0 \tag{71}$$

$$\Delta y^h = \sum_{n=0}^{n_{max}} \sum_{m=0}^{m_{max}} \frac{A_{mn} \cosh\left(t\sqrt{\frac{f}{a}}\right)}{an^2 + cm^2 + f} \ . \tag{72}$$

Combining the homogeneous solution (Eq. 72) and particular solution (Eq. 63) results in

$$\Delta y(s,t) = \sum_{n=0}^{n_{max}} \sum_{m=0}^{m_{max}} \frac{A_{mn}[\cosh\left(t\sqrt{\frac{f}{a}}\right) - \cos(ns)\cos(mt)] - C_{mn}\sin(ns)\cos(mt)}{an^2 + cm^2 + f} \ . \tag{73}$$

## Numerical Results

### a) Two-dimensional airfoils

The high non-linearity of flow-field governing equations, such as Euler and Navier-Stokes equations, has been suspected to cause significant reduction in the speed of MGM convergence [9]. To clarify this issue, three flow-field analysis codes have been used in conjunction with the original MGM and the Fourier series method. The three flow-field analysis codes were: a surface panel code with a compressibility correction, an Euler equation solution compressible flow code, and a compressible viscous flow Navier-Stokes code with a Baldwin-Lomax turbulence model.

Two airfoil design cases were examined. The first case utilized a target coefficient of pressure distribution corresponding to a NACA0012 non-lifting airfoil at free stream Mach number M = 0.5. The initial geometry in this case was a NACA 0009 non-lifting airfoil. The second case utilized the coefficient of pressure distribution of a NACA1311 cambered airfoil as its target. The initial geometry here was a NACA0012 non-lifting airfoil. In both cases, the inverse design process was performed using each of the three flow-field analysis codes in conjunction with the MGM procedure and the Fourier series technique. Composite plots of the convergence histories of these processes are shown in Fig. 3. For both design cases and all three flow-field analysis codes, the Fourier technique outperformed the MGM technique. A typical geometry and coefficient of pressure distribution evolution using the Fourier technique are shown in Fig. 4.

The slow convergence of the classical MGM technique with nonlinear flow-field analysis codes is most evident in the case of the lifting airfoil design (Fig. 3b). In this case, the Euler + MGM combination ceases to converge after three design iterations, and the Navier-Stokes + MGM combination ceases its convergence after five design iterations. This difficulty was not encountered when using the Fourier series technique, as in both the lifting and non-lifting cases the design converged faster with the Euler and Navier-Stokes analysis codes than with the panel code [11].

## b) Three-dimensional wings

Several wing design test cases were carried out using a three-dimensional panel code with an algebraic compressibility correction. The wingtip airfoil in each test case had a non-zero thickness. Each inverse shape design test case was performed with free-stream Mach number M = 0.2. For all three-dimensional results, the maximum number of Fourier terms to be considered in the s- and t-directions, $n_{max}$ and $m_{max}$, were set to 120.

The first panel case tested the Fourier technique's ability to modify a wing's thickness without affecting its symmetry. The wing planform was rectangular with a semi-span three times as large as the chord. The target pressure distribution was obtained from a three-dimensional panel code analysis of a wing whose airfoil shape varied smoothly from a NACA0012 airfoil at the root section to a NACA0009 airfoil at the wingtip. The initial guess wing had a NACA0012 airfoil shape at all span stations. The wing grid had 64 panels in the s-direction and 19 panels in the spanwise direction. The shape evolution parameters a, c, and f were set to 6.0, 0.5, and 1.5, respectively.

As shown in Figure 5, the designed wing's coefficient of pressure distribution nearly duplicated the target after ten calls to the flow-field analysis code. Small discrepancies in the pressure distribution can be seen at the wing tip section, and there is a noticeable bump in the pressure distribution of each section of the designed wing near the trailing edge. This trailing edge pressure spike, though physically unexplainable, occurred in each of the three-dimensional panel code tests that were performed. The shapes of the initial wing, the final wing, and the wing after one design iteration are shown in Fig. 6. The essential three-dimensionality of the shape change can be most notably seen at 0% span. Here, where the initial and the target airfoil geometries and corresponding pressure distributions are identical, one would expect no change of the root airfoil shape during the inverse shape design process. However, as the general trend of the entire wing is toward a lower thickness, the root section experiences some loss of thickness after one iteration. The root section must then regain its original shape. Thus, in such a case where the root section coefficient of pressure is unmodified in the design target, it could be quite useful if the entire root section of the wing geometry is kept unchanged. Also, at 30% span, the initial geometry modification overshoots the appropriate thickness of the wing, but then converges to the target shape after approximately ten design iterations.

The second panel test case examined wing twist. The wing planform was rectangular with a semi-span of two chord lengths. The target pressure distribution was calculated from analysis of a wing with NACA0012 airfoil with a one degree angle of attack at the root section and three degrees angle of attack at the tip. The initial guess geometry also had a NACA0012 airfoil shape, but included no wing twist. The wing grid had 64 panels in the s-direction and 14 panels in the spanwise direction, clustered toward the wingtip. Parameters a, c, and f were set to 6.0, 0.5, and 1.5 respectively. Figure 7 depicts the initial coefficient of pressure distributions at several span stations and their values after the tenth iteration with the Fourier series design method, and the target values. Again, a spike in the pressure distribution is being generated at the trailing edge of the wing, most noticeably at 80% of the wing span. Figure 8 depicts the initial geometry and the geometry after the first and the tenth iteration.

The third panel case involved evolving wing thickness, camber, and twist. The wing had a taper ratio of 0.5 and a leading edge backward sweep angle of 7.125 degrees. The semi-span was equal to two root chord lengths. The target pressure distribution was obtained from analysis of a wing having a NACA1311 root airfoil shape and a NACA2412 tip airfoil shape with three degrees angle of attack at the tip. The initial geometry had a constant NACA0009 airfoil shape with a negative one degree angle of attack at the tip. The wing grid had 64 panels in the s-direction and 19 panels in the spanwise direction that were clustered toward the wingtip. The shape evolution parameters a, c, and f were set to 7.0, 0.9, and 1.2. As seen in Fig. 9, the target pressure differs from the tenth iteration design values mostly near the wingtip and at the trailing edge. The evolution of the geometry is shown in Fig. 10.

The fourth test case was examined in conjunction with both a three-dimensional Euler equation solver and a turbulent Navier-Stokes equation solver. In these cases, a wingtip that quickly shrinks the airfoil thickness down to zero was added to each wing. The computational grid was regenerated after each application of the Fourier series design method by stacking two-dimensional C-grids generated for each span station. Additional grid layers were provided beyond the wingtip so that finite wing effects could be included.

The subsonic Euler design case was applied to a wing with a taper ratio of 0.5. The leading edge sweep angle was 14.03 degrees and the trailing edge had zero sweep, while the semi-span was two times the root chord length. The free stream Mach number was $M = 0.6$. The target pressure distribution corresponded to a severely twisted wing with a root airfoil NACA0009 at +4 degrees angle of attack and a tip airfoil NACA1311 at –4 degrees angle of attack. The initial geometry for the design process had root airfoil NACA2412 at –4 degrees angle of attack and a tip airfoil NACA0009 at +4 degrees angle of attack. Twenty span stations, 32 C-layers, 64 grid cells on each airfoil, and 16 cells along the wake defined the computational grid. The shape evolution parameters a, c, and f were set to 7.0, 0.9, and 1.2, respectively. The preliminary results indicate (Fig. 11) that target pressure distribution is not fully achieved near the root and tip after twenty design iterations. Figure 12 depicts the change in geometry of the wing during the design process.

The subsonic Navier-Stokes design case was identical to the subsonic Euler case, except that viscosity effects were included in the target pressure distribution. A Reynolds number of one million was used. The shape evolution parameters a, c, and f were set to 7.0, 0.9, and 1.2, respectively. The results of this inverse shape design case are shown in Figs. 13 and 14. Comparison with the results from the corresponding Euler case (Figs. 11 and 12) shows little difference in the performance of the code.

One Euler equation design case was attempted in transonic flight conditions, seeking to design a fully subsonic wing from a wing with a shock wave at the flight Mach number M = 0.8. The wing planform had a taper ratio of 0.5, leading edge sweep angle of 14.03 degrees, zero trailing edge sweep, and semi-span of two times the root chord length. The initial guess had a NACA0012 airfoil shape at 5 degrees angle of attack. The target pressure distribution corresponded to a high subsonic non-lifting wing with a NACA0009 airfoil. The computational grid was defined by 20 span stations, 32 C-layers, and 100 grid cells per airfoil. The shape evolution parameters a, c, and f were set to 7.0, 1.2, and 1.6, respectively. As shown in Fig. 15, after twenty iterations, the wing still had some lift despite its non-lifting target pressure distribution. More dramatic was the airfoil shape obtained after the first iteration of the Fourier series technique. A large dent was developed at the location of the shock wave on the upper surface of the wing (Fig. 16). A potential danger in supplying a discontinuous pressure distribution as the target pressure in the Fourier series method is that the surface curvature can locally overreact to the discontinuity, causing a concavity on the surface of the wing. However, after 20 iterations in this case, the concavity was removed, leaving a smooth wing shape (Fig. 16) and no shock wave (Fig. 17).


## Conclusions and Recommendations

A general formulation for the elastic membrane concept in aerodynamic shape inverse design has been explained. Details were given for a finite-difference based MGM method and for a Fourier series based analytical method for its implementation. The main advantage of both design methods is that they can be very easily programmed and used in conjunction with any available flow-field analysis code without a need for modifying such a code. From a detailed sequence of numerical tests of the Fourier series technique as applied to isolated airfoil design, it can be concluded that the convergence rate of this method does not depend on the non-linearity of the flow-field solver used. The Fourier series method was found to consistently converge faster than the MGM method, resulting in fewer calls to the time consuming flow-field analysis code. When applied to three-dimensional wing design, preliminary results show that the Fourier series technique is able to design subsonic wings in conjunction with a panel code, an Euler solver, or a turbulent Navier-Stokes solver. Cases involving shocked initial or target pressure distributions may be subject to overly dramatic changes in curvature in the region of the shock when using the simplified formulation presented here. It is possible that the inclusion of the mixed second partial derivative term and/or the first derivative terms on the left hand side of the elastic membrane shape evolution equation (56) could eliminate this problem by rapidly damping the elastic surface oscillations.

The addition of derivatives of pressure on the right hand side of the general elastic membrane model (Eqs. 45 and 46) could potentially significantly increase the convergence rate of the design process. Another possibility to significantly reduce the number of calls to the flow-field analysis code would be to devise a procedure for optimizing the user specified coefficients in the elastic membrane model equation.

## Acknowledgments

The authors would like to express their gratitude for the NASA Graduate Student Research Program Fellowship facilitated and monitored by Dr. John Malone, the National Science Foundation Grant DMI-9522854 monitored by Dr. George A. Hazelrigg, the NASA Lewis Research Center Grant NAG3-1995 facilitated by Dr. John K. Lytle and supervised by Dr. Kestutis Civinskas, and for Lockheed Martin Skunk Works research grant facilitated by Mr. Thomas Oatway and monitored by Dr. Anthony Thornton.

## References

[1]    Dulikravich, G.S.: Aerodynamic Shape Design and Optimization: Status and Trends, AIAA  Journal of Aircraft, Vol. 29, No. 5, pp. 1020-1026 (Nov./Dec. 1992).

[2]    Dulikravich, G.S.: Shape Inverse Design and Optimization for Three-Dimensional Aerodynamics, AIAA invited paper 95-0695, AIAA Aerospace Sciences Meeting, Reno, NV, January 9-12 (1995).

[3]    Dulikravich, G.S.: Design and Optimization Tools Development, chapters no. 10-15 in New Design Concepts for High Speed Air Transport, (editor: H. Sobieczky), Springer, Wien/New York, 1997, pp. 159-236 (1997).

[4]    Garabedian, P. and McFadden, G.: Computational Fluid dynamics of Airfoils and Wings, in Transonic, Shock, and Multidimensional Flows: Advances in Scientific Computing, pp. 1-16, Academic Press (1982).

[5]    Garabedian, P. and McFadden, G.: Design of Supercritical Swept Wings, AIAA J., Vol.20, No.3, pp. 289-291 (March 1982).

[6]    Malone, J., Vadyak, J. and Sankar, L.N.: A Technique for the Inverse Aerodynamic Design of Nacelles and Wing Configurations, AIAA Journal of Aircraft, Vol. 24, No. 1, pp. 8-9, (January 1987).

[7]    Hazarika, N.: An Efficient Inverse Method for the Design of Blended Wing-Body Configurations, Ph.D. Thesis, Aerospace Eng. Dept., Georgia Institute of Technology (June 1988).

[8]    Malone, J.B., Narramore, J.C. and Sankar, L.N.: An Efficient Airfoil Design Method Using the Navier-Stokes Equations, in Proceedings of AGARD Specialists' Meeting on Computational Methods for Aerodynamic Design (Inverse) and Optimization, AGARD-CP-463, (editor: J. Slooff), Loen, Norway, May 22-23 (1989).

[9]    Malone, J.B., Narramore, J.C. and Sankar, L.N.: An Airfoil Design Method for Viscous Flows, Proceedings of the $15^{th}$ Southeastern Conference on Theoretical and Applied Mechanics, Vol. XV, pp. 463-470, (editors: S.V. Hanagud, M.P. Kamat, C.E. Ueng), Georgia Institute of Technology, Atlanta, GA (1990).

[10]   Press, W.H, Teukolsky, S.A., Vetterling, W.T. and Flannery, B.P.:  Numerical Recipes in FORTRAN, The Art of Scientific Computing, 2nd Edition, Cambridge University Press, Cambridge (1986).

[11]   Dulikravich, G.S. and Baker, D.P.: Fourier Series Analytical Solution for Inverse Design of Aerodynamic Shapes, Proceedings of International Symposium on Inverse Problems in Engineering Mechanics – ISIP'98 (editors: M. Tanaka and G.S. Dulikravich), Nagano City, Japan, March 24-27, 1998, Elsevier Science, U.K. (1988).

[12]   Dulikravich, G.S. and Baker, D.P.: Aerodynamic Shape Inverse Design Using a Fourier Series Method, AIAA paper 99-0185, Aerospace Sciences Meeting, Reno, NV, January 11-14 (1999).

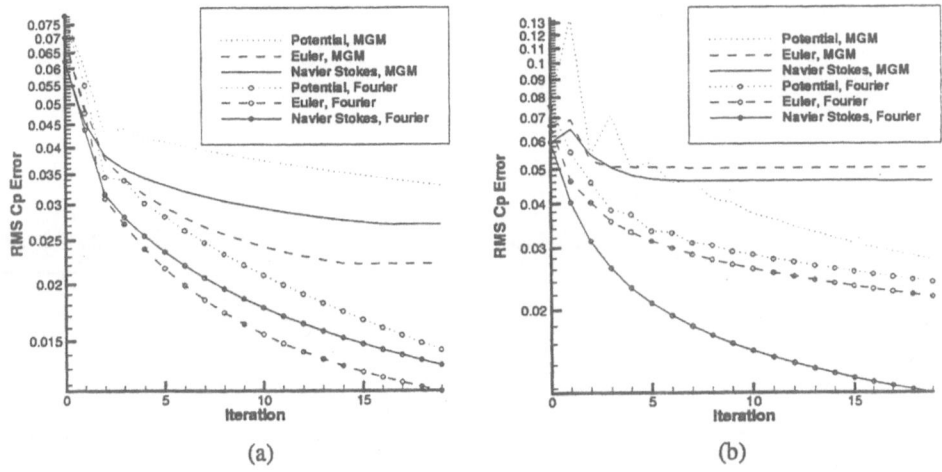

**Fig. 3: Convergence histories of MGM and Fourier methods applied to identical problems. M=0.5. Re=1000000 for Navier Stokes case.**
**(a) Nonlifting case: NACA 0009 evolves into NACA 0012. Beta=(1.2, 0.0, 0.4).**
**(b) Lifting case: NACA 0012 evolves into NACA 1311. Beta=(1.4, 0.0, 0.6).**

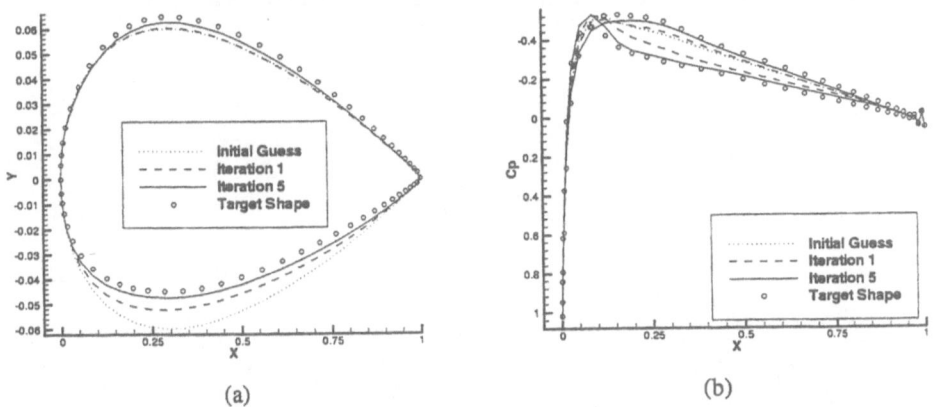

**Fig. 4: Fourier method: evolution from NACA 0012 to NACA 1311 airfoil. Navier Stokes solver. M=0.5. Re=1000000. Beta=(1.4, 0.0, 0.6)**
**(a) Evolution of geometry. Y-axis enlarged for clarity.**
**(b) Evolution of surface coefficient of pressure.**

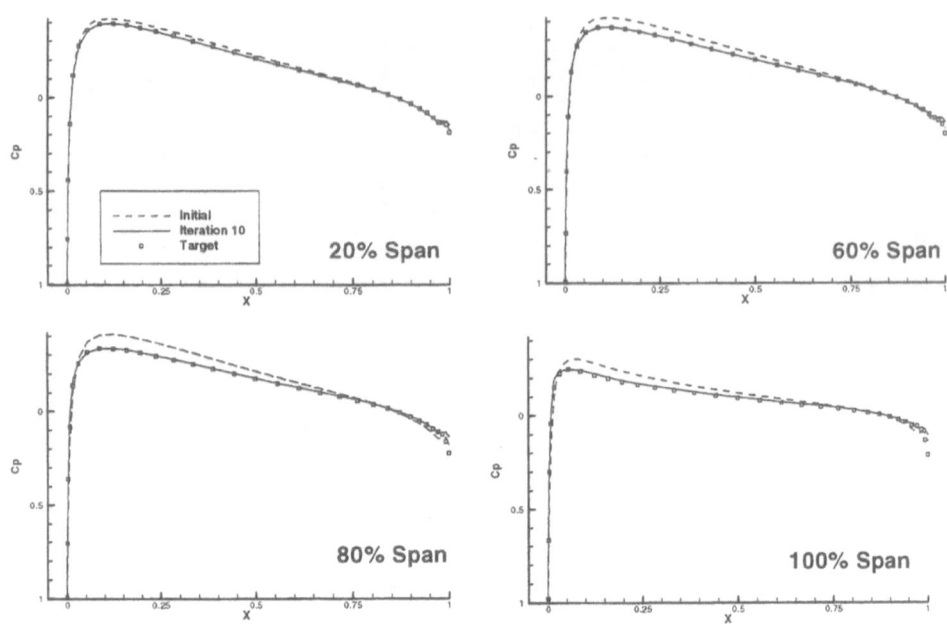

**Fig. 5: Sectionwise Cp distributions for panel code symmetric case no. 1**

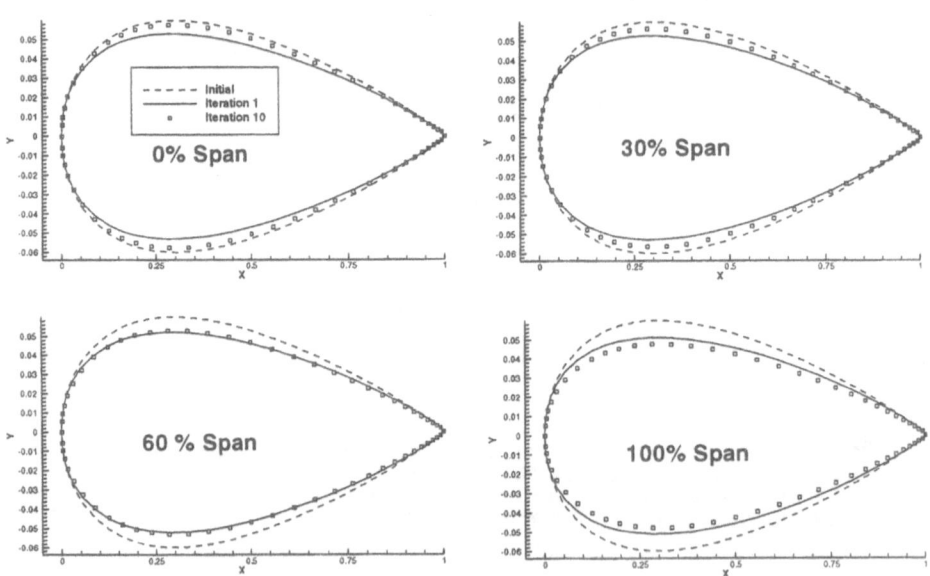

**Fig. 6: Geometry evolution by section for symmetric panel code symmetric case no. 1. Y-axis enlarged for clarity.**

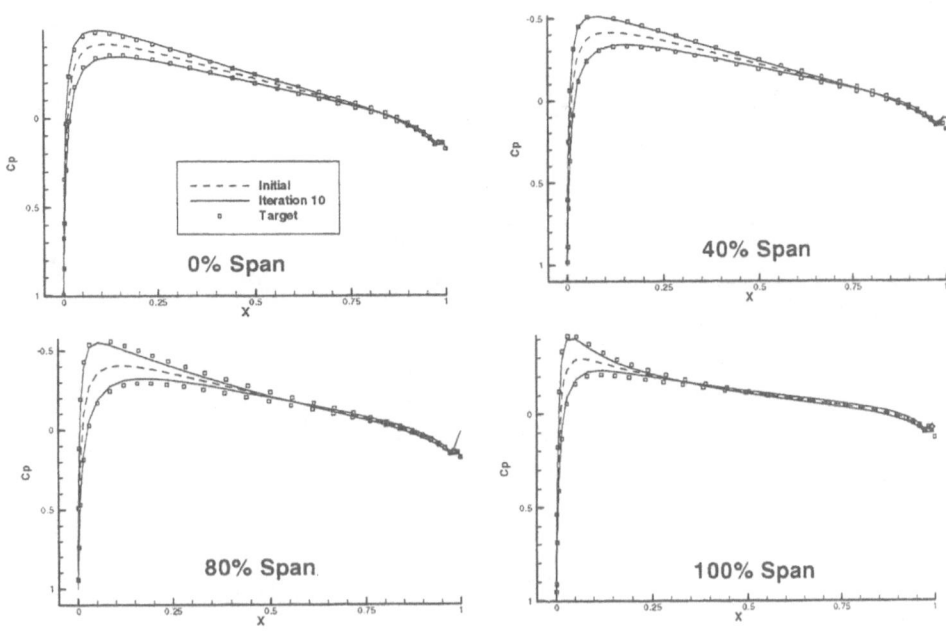

**Fig. 7:** Sectionwise Cp distributions for panel code twisted case no. 2.

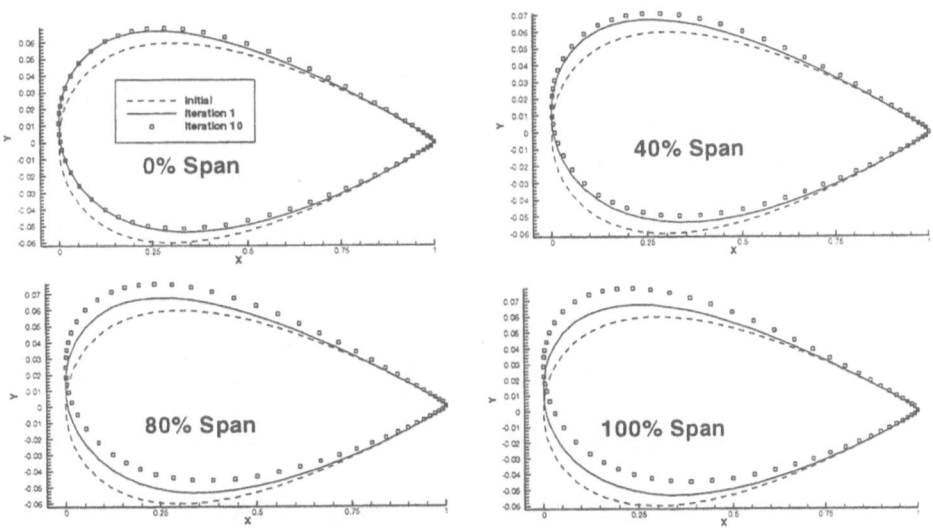

**Fig. 8:** Geometry evolution by section for panel code twisted case no. 2.
Y-axis enlarged for clarity.

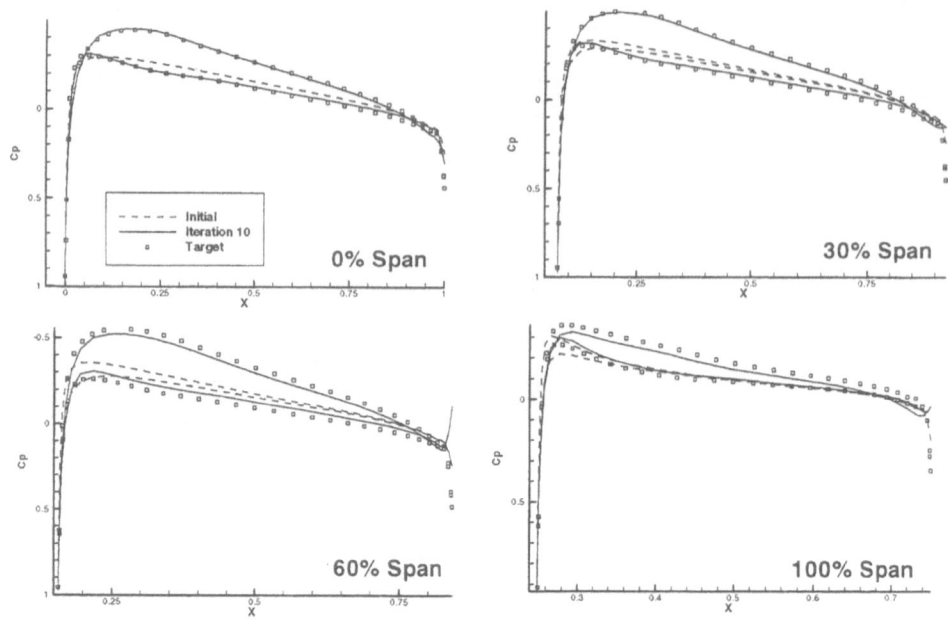

**Fig. 9: Sectionwise Cp distributions for panel code lifting case no. 3.**

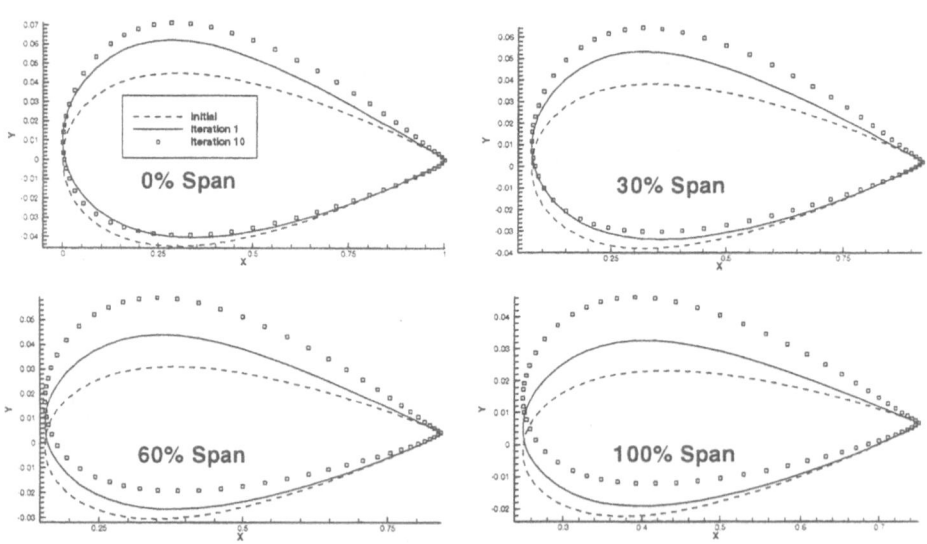

**Fig. 10: Geometry evolution by section for panel code lifting case no. 3.
Y-axis enlarged for clarity.**

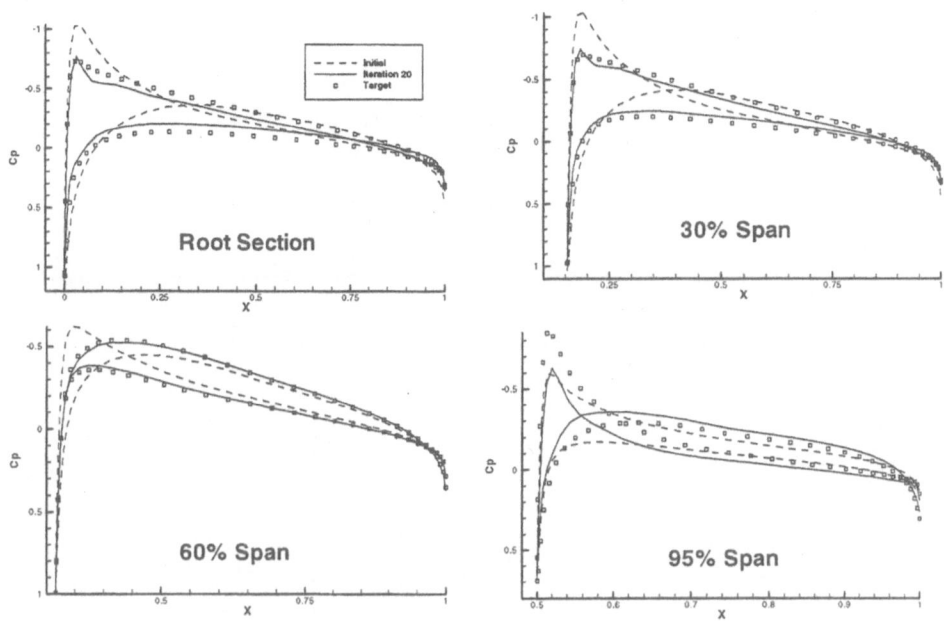

**Fig. 11:** Sectionwise Cp distributions for severely twisted Euler subsonic case.

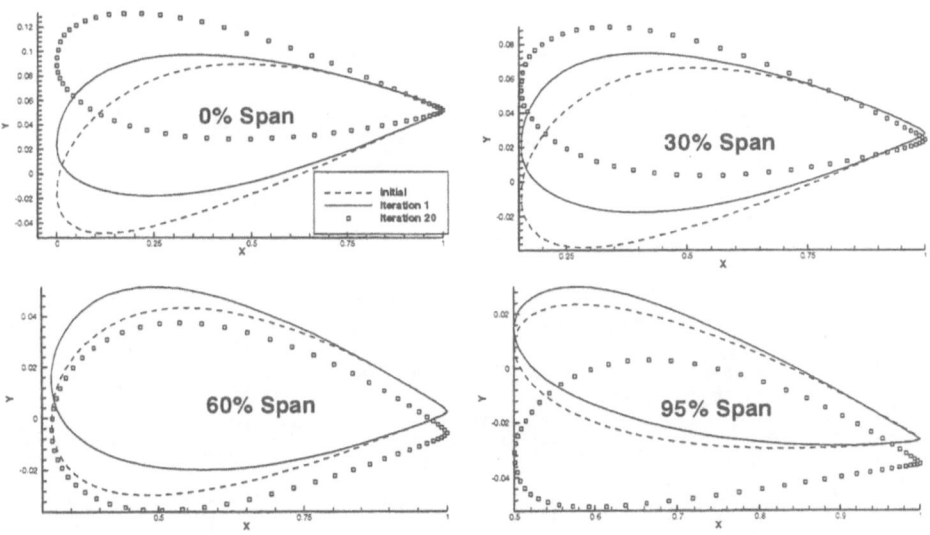

**Fig. 12:** Geometry evolution by section for severely twisted Euler subsonic case. Y-axis enlarged for clarity.

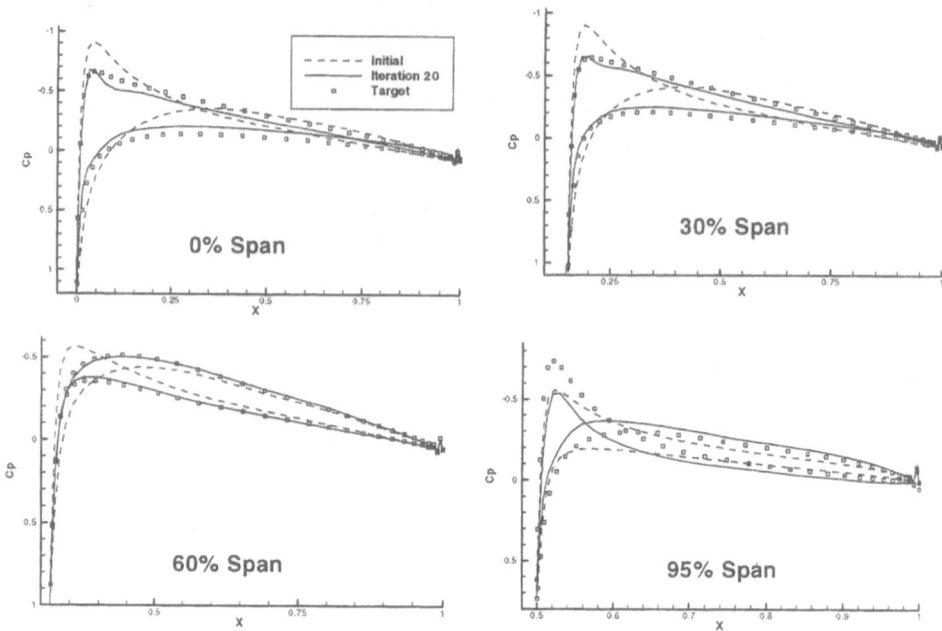

**Fig. 13: Sectionwise Cp distributions for severely twisted Navier-Stokes subsonic case.**

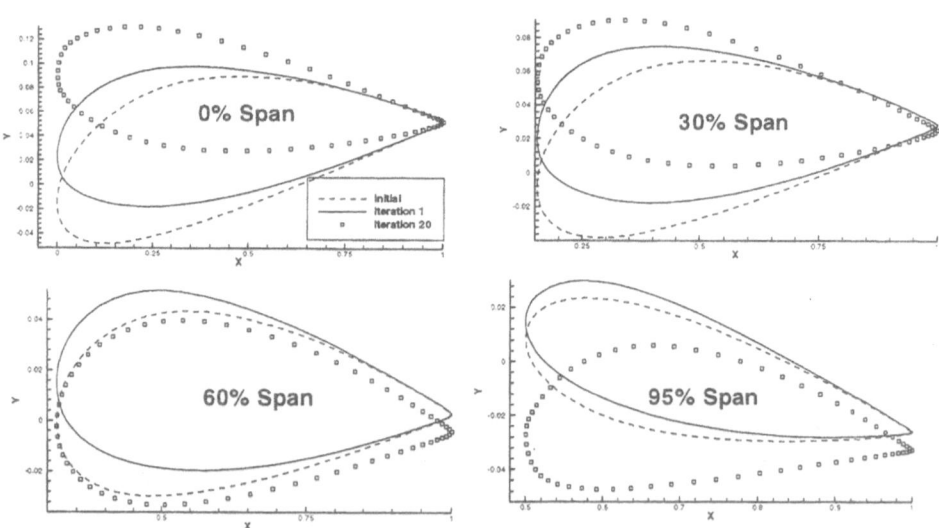

**Fig. 14: Geometry evolution by section for severely twisted Navier-Stokes subsonic case. Y-axis enlarged for clarity.**

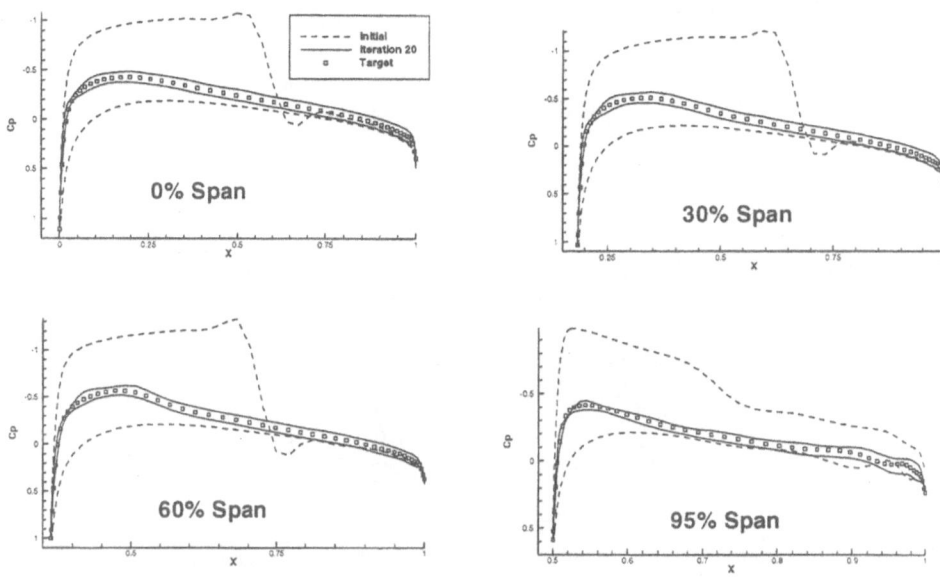

**Fig. 15:** Sectionwise Cp distributions for Euler shock removal case.

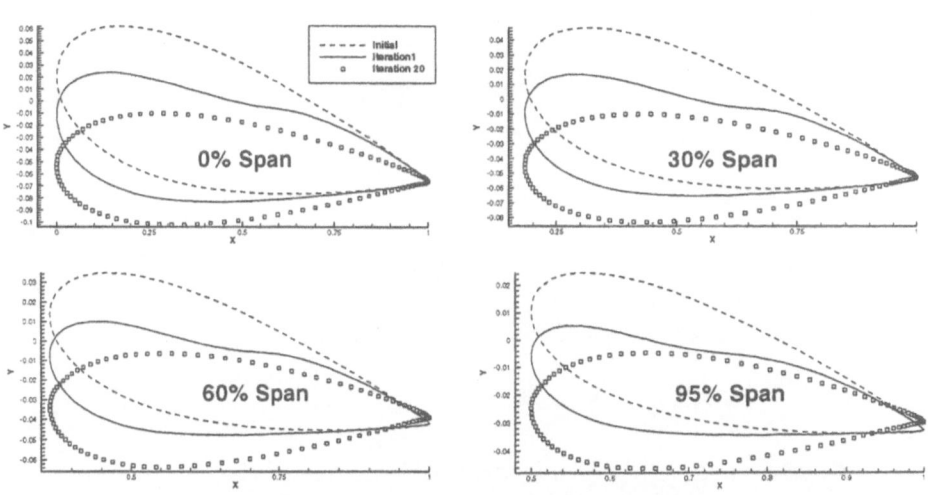

**Fig. 16:** Geometry evolution by section for Euler shock removal case.
Y-axis enlarged for clarity

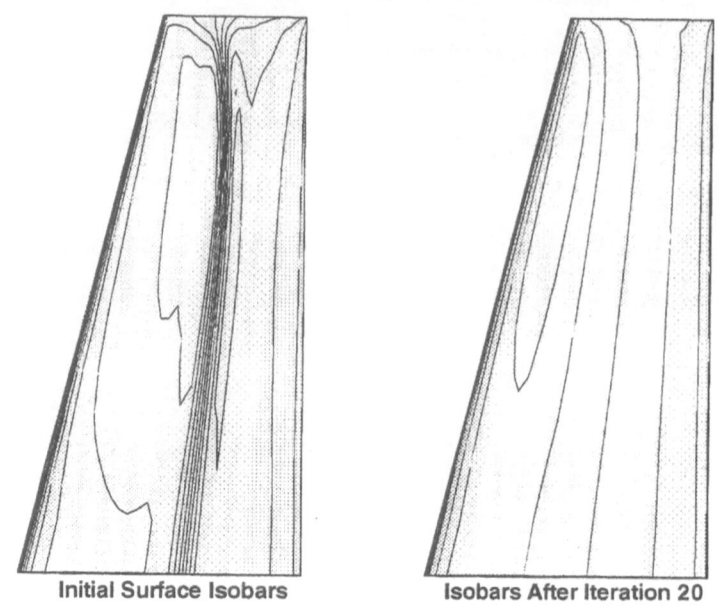

Initial Surface Isobars        Isobars After Iteration 20

**Fig. 17: Cp contours on upper side of wing for Euler shock removal case.**

# Numerical Methods for Inverse Solution in Aerodynamic Design of Turbomachinery

Naixing Chen

Institute of Engineering Thermophysis
Chinese Academy of Sciences
P.O. Box 2706, Beijing 100080, China
Tel: (8610)-62561435. Fax: (8610)-62575913
E-mail: nxc@etpserver.etp.ac.cn

## Summary

A survey is given on aerodynamic inverse solution activities by the authors in the Institute of Engineering Thermophysics. The survey includes stream function(SF) method of two different kinds of stream surfaces, which are usually assumed for the design in turbomachinery, stream-function-coordinate(SFC) method for two-dimensional and three-dimensional flows, potential function(PF) method and the method based on the direct solution with a residual-correction technique(DS-RC). The application of these methods and computer codes to turbine and compressor design has shown that the use of inverse solution techniques enables engineers to improve the design. A set of calculation examples and figures is also illustrated in the paper.

## 1. Introduction

The need for highly efficient turbomachinery has become clear in the past decade due to rising fuel costs. At same time, the costs of large-scale testing of new designs have also escalated, making preliminary design using computational methods more attractive. Utilizing the aerodynamic optimization problem of blade design and boundary layer theory, the optimal velocity distribution is obtained. This is on the one hand. On the other hand, the blade geometry can be estimated by fluid mechanics theory, such as conform mapping method, singularity method, and various methods based on using stream function, potential function or their combinations. Due to the rapid progress of computer science and computational techniques, there is an opportunity to deal with numerical computation for three-dimensional turbulent flows in turbomachinery. It gives a possibility to develop inverse solution method for viscous flow.

Many design procedures, including aerodynamic design of turbomachinery, are mostly based on the repeated application of a direct solver, and therefore iterative and indirect design methods. In the case of direct problem, prediction of the details of a flow field is asked if the geometry of the object is specified. Usually, using a direct solver, it is difficult to obtain a geometry that gives exactly the target surface pressure (or velocity) or blade loading distributions. Also, the computational cost may be prohibitively high in these iterative procedures since the flow solutions must be repeated several times until the geometry converges. Due to these limitations there is a continuing interest in developing numerical methods to solve the inverse and hybrid problems. In the case of inverse or design problem, prediction of the geometry of the object, which must be compatible with the desired features of the flow field, is required.

Hybrid solution is a kind of inverse solution problems. For this case, the geometry of a portion of the object is unknown, and the remainder is determined by a given prescribed velocity distribution. It is useful to use an inverse solver for obtaining an aerodynamically feature that leads to a realistic geometry. As indicated in [1, 2], for two-dimensional arbitrarily described velocity distributions on suction and pressure surfaces, a solution exits when a circulation constraint is satisfied.

It is well known that the formulation of an optimal velocity distribution on the suction surface requires a constraint to avoid boundary layer separation [3]. Cascade wind tunnel experiments and calculations have shown that the aerodynamic conditions on the suction surface play a more important role in the overall blade loss than those on the pressure surface. This leads to a very careful design of the suction surface of a blade, which can be obtained by the hybrid problem solution.

In the present paper the following contents are emphasized on: stream function (SF) method, stream-function-coordinate(SFC) method, potential function(PF) method and a direct solution method with a residual-correction technique(DS-RC). The latter method is based on a Navier -Stokes solver.

## 2. Governing Equations

The relative flow in a turbomachine is assumed to be steady and inviscid. The governing equations expressed in vector form are as follows:

*Continuity equation:*

$$\nabla \cdot \left( \rho \vec{W} \right) = 0 .\tag{2-1}$$

*Momentum equation:*

$$-\vec{W} \times \left( \nabla \times \vec{W} \right) + 2\vec{\Omega} \times \vec{W} = -\nabla I + T \nabla S .\tag{2-2}$$

*Formula for calculating density for isentropic process:*

$$\rho = \rho_1 \left( T/T_1 \right)^{1/(k-1)} \Big/ \exp\!\left[ \left( s - s_1 \right)/R \right] .\tag{2-3}$$

*The definition of rothalpy:*

$$I = C_p T + \frac{W^2}{2} - \frac{(\Omega r)^2}{2} .\tag{2-4}$$

*The equation of state of the prefect gas:*

$$p = \rho R T .\tag{2-5}$$

In the above equations, $\vec{W}$ and $\vec{\Omega}$ represent the relative velocity vector of fluid particles and the angular velocity vector of rotation, respectively, and the relationship between absolute and relative velocity vectors is: $\vec{V} = \vec{W} + \vec{\Omega} \times \vec{r}$. $\rho, p, T, I$ and $s$ indicate density, pressure, absolute temperature, rothalpy and entropy, respectively. Subscript 1 denotes a certain state of the thermodynamic process.

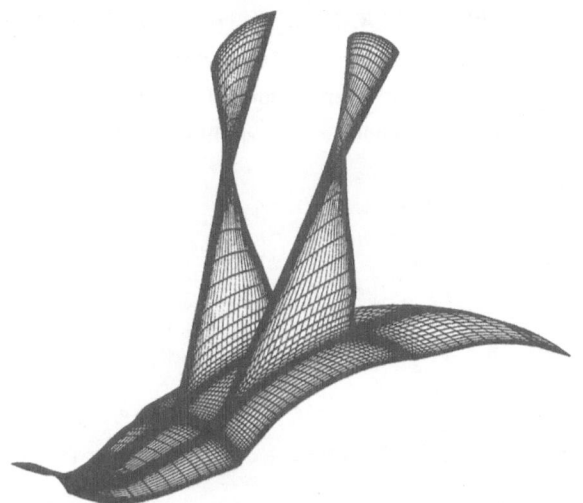

Fig.2-1. A single rotor compressor and its computational grid system

Due to the complexity of geometry it is convenient to express the three-dimensional flow by using body-fitted curvilinear coordinate system. For example, Fig.2-1 demonstrates a single rotor compressor and its coordinate systems. In the figure only two blades are demonstrated. In the following section we can see that the use of the body-fitted non-orthogonal curvilinear coordinate system enables to derive an inverse stream-function-method.

The *continuity equation* and the *momentum equations* in three directions can also be written in the following form:

$$\frac{\partial\left(\rho\sqrt{g}w^i\right)}{\partial x^i} = 0 \quad (i = 1,2,3) \tag{2-6}$$

$\vec{e}^1$-direction
$$-w^2\left(\frac{\partial w_2}{\partial x^1} - \frac{\partial w_1}{\partial x^2}\right) + w^3\left(\frac{\partial w_1}{\partial x^3} - \frac{\partial w_3}{\partial x^1}\right) + 2\sqrt{g}\left(\omega^2 w^3 - \omega^3 w^2\right) = -\frac{\partial I}{\partial x^1} + T\frac{\partial s}{\partial x^1} \tag{2-7}$$

$\vec{e}^2$-direction
$$-w^3\left(\frac{\partial w_3}{\partial x^2} - \frac{\partial w_2}{\partial x^3}\right) + w^1\left(\frac{\partial w_2}{\partial x^1} - \frac{\partial w_1}{\partial x^2}\right) + 2\sqrt{g}\left(\omega^3 w^1 - \omega^1 w^3\right) = -\frac{\partial I}{\partial x^2} + T\frac{\partial s}{\partial x^2} \tag{2-8}$$

$\vec{e}^3$-direction
$$-w^1\left(\frac{\partial w_1}{\partial x^3} - \frac{\partial w_3}{\partial x^1}\right) + w^2\left(\frac{\partial w_3}{\partial x^2} - \frac{\partial w_2}{\partial x^3}\right) + 2\sqrt{g}\left(\omega^1 w^2 - \omega^2 w^1\right) = -\frac{\partial I}{\partial x^3} + T\frac{\partial s}{\partial x^3} \tag{2-9}$$

where $\omega^i$ represents the contravariant component of the annular velocity vector of rotation $\vec{\Omega}$, while $g$ represents the metric tensor determinant. $w_i, w^i$ denote the covariant and the contravariant components of the velocity vector $\vec{W}$  The relative velocity vector can be expressed by the following expressions as:

$$\vec{W} = W^1\vec{u}_1 + W^2\vec{u}_2 + W^3\vec{u}_3 = w^1\vec{e}_1 + w^2\vec{e}_2 + w^3\vec{e}_3, \qquad (2\text{-}10a)$$

$$\vec{W} = W_1\vec{u}^1 + W_2\vec{u}^2 + W_3\vec{u}^3 = w_1\vec{e}^1 + w_2\vec{e}^2 + w_3\vec{e}^3 \qquad (2\text{-}10b)$$

where $\vec{u}_1, \vec{u}_2$ and $\vec{u}_3$ are unit vectors; $\vec{u}^1, \vec{u}^2$ and $\vec{u}^3$ are reciprocal unit vectors; $\vec{e}_1, \vec{e}_2$ and $\vec{e}_3$ are base vectors; $\vec{e}^1, \vec{e}^2$ and $\vec{e}^3$ are reciprocal base vectors.

Later we will use these equations to develop inverse solution methods for turbomachine flow.

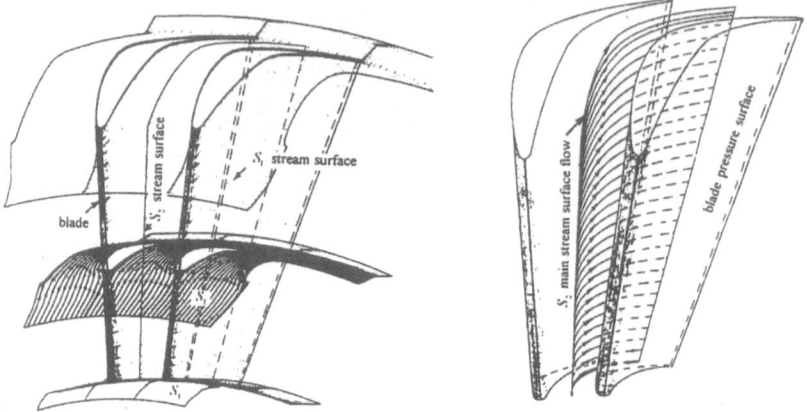

Fig.3-1. A quasi-3D flow model of a turbine stator with a main $S_2$ stream surface and several $S_1$ stream surfaces (in the figure only three are shown)

## 3. Stream Function Method

Wu in his classical work[4] proposed firstly that the three-dimensional flow in turbomachinery can be solved by alternatively successive iterations of two kinds of stream surfaces. He suggested that the governing equations of the flows on these two kinds of stream surfaces are expressed in form of stream function equations. Then, for the quasi-three-dimensional flow the solution is carried on only for one meridional main $S_2$ stream surface and several $S_1$ stream surfaces of revolution. In this case the solution of the $S_2$ main stream surface flow is also a kind of inverse problems. It is solved by a target parameter distribution, such as circulation, flow angle, specific flow rate, etc. In the paper of [5] these two kinds of stream function equations are expressed by non-orthogonal coordinate system and non-orthogonal velocity components.

Many authors (Liu[7], Chen[1,2], Ge[8] and Wang[9]) developed their inverse methods by using stream function equation. It is very interesting that Luu et al[10,11] have solved an inverse problem of quasi-three-dimensional turbomachine flow using $S_1 - S_2$ approach.

In the design stage, the concept based on quasi-three-dimensional model with a main meridional $S_2$ stream surface and several $S_1$ stream surface flows of revolution is usually used(see Fig.3-1). The left figure of Fig.3-1 shows that the main $S_2$ stream surface is located in the channel between two blades. In Fig.3-1 there is a quasi-three-dimensional flow model. In the figure only three stream surfaces of revolution are demonstrated. The aero-thermodynamic parameters on this main meridional $S_2$ stream surface are approximately equal to the circumferentially averaged parameters. In this section the following stream function equations of the $S_1$ and $S_2$ flows are discussed. In the quasi-three-dimensional procedure usually $S_2$ is firstly calculated than $S_1$. Therefore, we will discuss the stream function equation of $S_2$ stream surface flow firstly. Later, when we will discuss the solution procedure of the stream function method(SF), we will emphasize on the solution of $S_1$ stream surface flow.

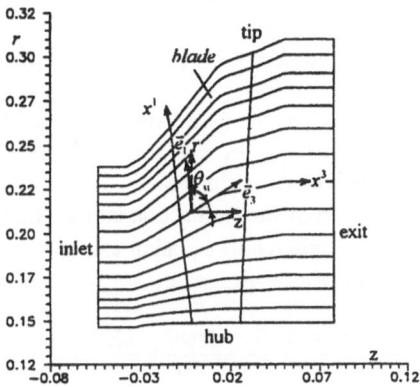

Fig.3-2. Meridional pattern of a turbine stator and the non-orthogonal coordinates

## 3.1 $S_2$ ( Meridional) Stream Surface Flow

Since the meridional pattern is geometrically complicated, usually the stream function equation adopted is expressed in non-orthogonal coordinate system. This system is body-fitted. It was already shown before. Let we assume that $x^2 = \varphi$; $x^3, x^2$ and $x^1$ are meridional, circumferencial and span directions, respectively; $x^3, x^1$ , lying on the meridianal plane, are the non-orthogonal coordinates. Then, the vectors of $x^3$ and $x^1$ (i.e. $\bar{e}_3$, $\bar{e}_1$ ) are perpendicular to the circumferential direction. Fig.3-3 shows a meridional pattern of a typical steam turbine stator. In the formula of calculating the metric tensor determinant of $\sqrt{g}$ (representing the cubic element of fluid) we should consider the thickness effect of the stream surface in the $\varphi$-direction, $\tau$, i.e. $\sqrt{g} = \tau \bar{e}_2 \bullet (\bar{e}_3 \times \bar{e}_1)$. The absolute value of $\tau$ does not have any practical meaning in the calculation, and it represents the circumferencial thickness effect of the blade channel. In this case it can be approximately written as $\tau = 1 - B(blade \ number \ / \ 2\pi r)$, where $B$ is blade thickness. Satisfying the continuity equation Eq.(2-6), we have the following expressions for defining the stream function as:

$$\bar{\partial}\psi / \bar{\partial}x^1 = \rho\sqrt{g}w^3 \ \ and \ \ \bar{\partial}\psi / \bar{\partial}x^3 = -\rho\sqrt{g}w^1. \tag{3-1}$$

where $\bar{\partial}(\ )/\partial x^3$, $\bar{\partial}(\ )/\partial x^1$ are the first partial derivatives on the stream surface, and they are defined as:

$$\frac{\bar{\partial}(\ )}{\partial x^1} = \frac{\partial(\ )}{\partial x^1} + \frac{n_1}{n_2}\frac{\sqrt{g_{11}}}{r}\cos(\bar{e}^1,\bar{e}_1)\frac{\partial(\ )}{\partial \varphi} \ ; \tag{3-2a}$$

$$\frac{\bar{\partial}(\ )}{\partial x^3} = \frac{\partial(\ )}{\partial x^3} + \frac{n_3}{n_2}\frac{\sqrt{g_{33}}}{r}\cos(\bar{e}^3,\bar{e}_3)\frac{\partial(\ )}{\partial \varphi} \tag{3-2b}$$

where $n_1, n_2$ and $n_3$ are the physical components of unit normal vector. $(\bar{e}^i,\bar{e}_i)$ is the angle between vectors of $\bar{e}^i$ and $\bar{e}_i$. The absolute velocity vector can be expressed as:

$$\vec{V} = v^1\bar{e}_1 + v^2\bar{e}_2 + v^3\bar{e}_3 \text{ or } \vec{V} = v^1\bar{e}_1 + [w^2 + (\vec{\Omega} \times \vec{r})^2]\bar{e}_2 + v^3\bar{e}_3 \ . \tag{3-3}$$

The combination of the first and the third terms is the meridional component of the absolute (and also the relative) velocity vector. $v^1, v^2$ and $v^3$ denote contravariant components of the absolute velocity vector on $\bar{e}_1$, $\bar{e}_2$ and $\bar{e}_3$ directions, respectively. Then, substituting the definition of stream function into the momentum equation in $\bar{e}_3$- direction, the stream function differential equation can be written in the form of non-orthogonal coordinate system as:

$$\frac{1}{g_{11}}\frac{\bar{\partial}^2\psi}{\bar{\partial}(x^1)^2} - \frac{2\cos\theta_{31}}{\sqrt{g_{11}g_{33}}}\frac{\bar{\partial}^2\psi}{\partial x^1\bar{\partial}x^3} + \frac{1}{g_{33}}\frac{\bar{\partial}^2\psi}{\bar{\partial}(x^3)^2} + \frac{E}{\sqrt{g_{11}}}\frac{\bar{\partial}\psi}{\partial x^1} + \frac{F}{\sqrt{g_{33}}}\frac{\bar{\partial}\psi}{\partial x^3} = G \ , \tag{3-4}$$

where the coefficients are as follows:

$$E = \frac{1}{\sqrt{g_{11}}}\frac{\bar{\partial}}{\partial x^1}(\ln\frac{\sqrt{g_{33}}}{\sqrt{g_{11}}\sin\theta_{31}}) - \frac{1}{\sqrt{g_{11}}}\frac{\bar{\partial}\ln\tau}{\partial x^1}$$
$$+ \cos\theta_{31}\frac{1}{\sqrt{g_{33}}}\frac{\bar{\partial}\ln\tau}{\bar{\partial}x^3} + \frac{1}{\sin\theta_{31}\sqrt{g_{33}}}\frac{\bar{\partial}\theta_{31}}{\bar{\partial}x^3} \tag{3-5a}$$

$$F = \frac{1}{\sqrt{g_{33}}}\frac{\bar{\partial}}{\partial x^3}(\ln\frac{\sqrt{g_{11}}}{\sqrt{g_{33}}\sin\theta_{31}}) - \frac{1}{\sqrt{g_{33}}}\frac{\bar{\partial}\ln\tau}{\bar{\partial}x^3}$$
$$+ \cos\theta_{31}\frac{1}{\sqrt{g_{11}}}\frac{\bar{\partial}\ln\tau}{\partial x^1} + \frac{1}{\sin\theta_{31}\sqrt{g_{11}}}\frac{\bar{\partial}\theta_{31}}{\partial x^1} \tag{3-5b}$$

$$G = \frac{1}{\sqrt{g_{11}}}\frac{\bar{\partial}\psi}{\partial x^1}(\frac{1}{\sqrt{g_{11}}}\frac{\bar{\partial}\ln\rho}{\partial x^1} - \frac{1}{\sqrt{g_{33}}}\frac{\bar{\partial}\ln\rho}{\bar{\partial}x^3}\cos\theta_{31}) + \frac{1}{\sqrt{g_{33}}}\frac{\bar{\partial}\psi}{\bar{\partial}x^3}(\frac{1}{\sqrt{g_{33}}}\frac{\bar{\partial}\ln\rho}{\bar{\partial}x^3}$$
$$- \frac{1}{\sqrt{g_{11}}}\frac{\bar{\partial}\ln\rho}{\partial x^1}\cos\theta_{31}) + \tau\sin\theta_{31}\rho H \ . \tag{3-5c}$$

$$H = (-\frac{W_\varphi}{r\sqrt{g_{11}}}\frac{\bar{\partial}(V_\varphi r)}{\partial x^1} + \frac{1}{\sqrt{g_{11}}}(\frac{\bar{\partial}I}{\partial x^1} - T\frac{\bar{\partial}s}{\partial x^1}) - F_{blade})/W^3. \qquad (3\text{-}5d)$$

In the above-mentioned formulas $\theta_{31}$ is the angle between the coordinates of $x^1$ and $x^3$; $W$ denotes the absolute value of relative velocity vector of $\bar{W}$, and $W^3$ is its physical contravariant component in $\bar{e}_3$ direction; $F_{blade}$ represents a blade force due to existing the circumferencial pressure gradient. It can be calculated by

$$F_{blade} = -\frac{1}{\rho r}\frac{\partial p}{\partial \varphi}\frac{n_1}{n_2}\sin\theta_{31} . \qquad (3\text{-}6)$$

For satisfying the stream surface condition the following expression can be obtained by

$$\bar{W}\cdot\bar{n} = 0 \qquad (3\text{-}7)$$

where $\bar{n}$ represents the unit normal vector to the stream surface.

For inverse problem solution there are various versions of the target parameters given. They are:

1.  circulation, $V_\varphi r = f_1(x^3, x^1)$;
2.  relative flow angle for relative flow (absolute flow angle for stator blade case), $\beta = f_2(x^3, x^1)$;
3.  specific flow rate, $\rho W_i = f_3(x^3, x^1)$ etc.

When the distribution of one of the target parameters is assigned on the plane, after solving the above-mentioned stream function equation the stream surface form can be obtained by integration of the following equation:

$$\varphi = \varphi_i + \int_{x_i^1}^{x^1}\frac{W^2}{W^1}\frac{\sqrt{g_{11}}}{r}dx^1 . \qquad (3\text{-}8)$$

From the above equations and formulas it is seen that the solution is an iterative process.

### 3.2 $S_1$ Stream Surface Flow

Here we assume that the $S_1$ stream surface to be a *surface of revolution* and a steady inviscid isentropic flow. Therefore, the span-wise pressure gradient and changes of rothalpy and entropy can be ignored. The coordinate system of this $S_1$ stream surface flow is adopted as $x^2, x^3$ of a non-orthogonal coordinate system, and $x^1$ is perpendicular to the plane of $x^2, x^3$ (see Fig.3-4 and Fig.3-5). Then, satisfying the continuity equation the definition of stream function is:

$$\partial\psi/\partial x^2 = \rho\sqrt{g}w^3 \quad \text{and} \quad \partial\psi/\partial x^3 = -\rho\sqrt{g}w^2. \qquad (3\text{-}9)$$

119

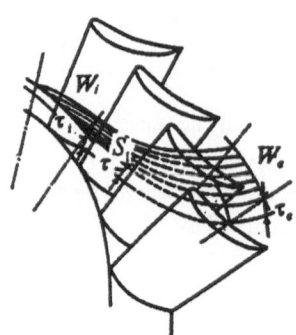

Fig.3-3. Principle scheme of $S_1$ stream
surface of revolution.

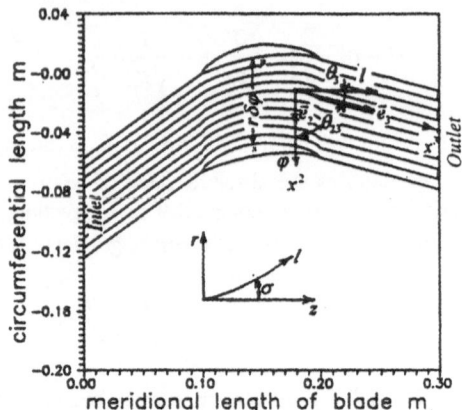

Fig.3-4. Non-orthogonal coordinates of a $S_1$ stream
surface of revolution

Substituting the definition of stream function into the momentum equation of $\bar{e}_2$-direction, the stream function differential equation can be written in the form of non-orthogonal coordinate system as:

$$\frac{1}{g_{22}}\frac{\bar{\partial}^2\psi}{\bar{\partial}(x^2)^2} - \frac{2\cos\theta_{23}}{\sqrt{g_{22}g_{33}}}\frac{\bar{\partial}^2\psi}{\bar{\partial}x^2\bar{\partial}x^3} + \frac{1}{g_{33}}\frac{\bar{\partial}^2\psi}{\bar{\partial}(x^3)^2} + \frac{E}{\sqrt{g_{22}}}\frac{\bar{\partial}\psi}{\bar{\partial}x^2} + \frac{F}{\sqrt{g_{33}}}\frac{\bar{\partial}\psi}{\bar{\partial}x^3} = G \ , \quad (3\text{-}10)$$

where the coefficients are as follows:

$$E = \frac{1}{\sqrt{g_{22}}}\frac{\bar{\partial}}{\bar{\partial}x^2}(\ln\frac{\sqrt{g_{33}}}{\sqrt{g_{22}}\sin\theta_{23}}) - \frac{1}{\sqrt{g_{22}}}\frac{\bar{\partial}\ln\tau}{\bar{\partial}x^2}$$
$$+ \cos\theta_{23}\frac{1}{\sqrt{g_{33}}}\frac{\bar{\partial}\ln\tau}{\bar{\partial}x^3} + \frac{1}{\sin\theta_{23}\sqrt{g_{33}}}\frac{\bar{\partial}\theta_{23}}{\bar{\partial}x^3} \qquad , \qquad (3\text{-}11a)$$

$$F = \frac{1}{\sqrt{g_{33}}}\frac{\bar{\partial}}{\bar{\partial}x^3}(\ln\frac{\sqrt{g_{22}}}{\sqrt{g_{33}}\sin\theta_{23}}) - \frac{1}{\sqrt{g_{33}}}\frac{\bar{\partial}\ln\tau}{\bar{\partial}x^3}$$
$$+ \cos\theta_{23}\frac{1}{\sqrt{g_{22}}}\frac{\bar{\partial}\ln\tau}{\bar{\partial}x^2} + \frac{1}{\sin\theta_{23}\sqrt{g_{22}}}\frac{\bar{\partial}\theta_{23}}{\bar{\partial}x^2} \qquad , \qquad (3\text{-}11b)$$

$$G = \frac{1}{\sqrt{g_{22}}}\frac{\bar{\partial}\psi}{\bar{\partial}x^2}(\frac{1}{\sqrt{g_{22}}}\frac{\bar{\partial}\ln\rho}{\bar{\partial}x^2} - \frac{1}{\sqrt{g_{33}}}\frac{\bar{\partial}\ln\rho}{\bar{\partial}x^3}\cos\theta_{23}) + \frac{1}{\sqrt{g_{33}}}\frac{\bar{\partial}\psi}{\bar{\partial}x^3}(\frac{1}{\sqrt{g_{33}}}\frac{\bar{\partial}\ln\rho}{\bar{\partial}x^3}$$
$$- \frac{1}{\sqrt{g_{22}}}\frac{\bar{\partial}\ln\rho}{\bar{\partial}x^2}\cos\theta_{23}) + 2\omega\rho\tau\sin\sigma\sin^2\theta_{23} \quad . \qquad (3\text{-}11c)$$

In the above-mentioned formulas $\theta_{23}$ is the angle between the coordinates of $x^2$ and $x^3$; $\tau$ is a stream surface thickness in the direction normal to the stream surface.

## 3.3 Solution Procedure

As pointed out in the previous section the solution procedure, we will mainly discuss, is the $S_1$ stream surface flow. In the present method the fluid flow is assumed to be an inviscid perfect gas. Fluid velocity is taken to be relative to a rotating reference frame which angular velocity relative to the inertial frame. Then, the governing equations, expressed on the two-dimensional stream surface of revolution by a non-orthogonal coordinate system, can be easily obtained from Eq.(3-1)- Eq.(3-10). In the latter formulas we will cancel the upper bar appeared on the top of the partial differential symbol of $\partial$. In stead of stream function we use a non-dimensional stream function, $\bar{\psi} = \psi / G_0$ . $G_0$ represents the flow rate past through a blade channel.

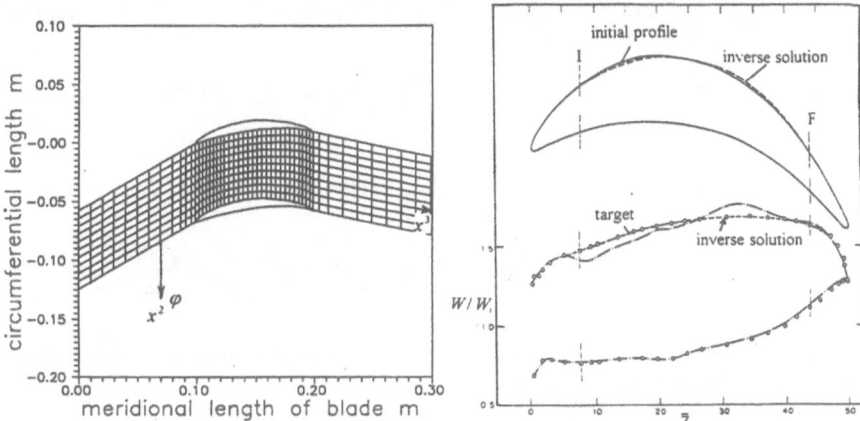

Fig.3-5. Grid system of $S_1$ computational domain.    Fig.3-6. A turbine cascade profile and its velocity distributions on blade surfaces

Now, we introduce a special kind of non-orthogonal coordinate system to the present method. It is forming the $x^3$- coordinate lines by keeping $(\varphi - \varphi_p) / \delta\varphi = const$ . $\varphi_p$ is the angular coordinate of the blade pressure surface; $\delta\varphi$ is the difference between the angular coordinates of suction and pressure surface, and it can be written as $\delta\varphi = pitch - blade\ thickness = \varphi_s - \varphi_p$ . Then we have the transformation formulas from cylindrical coordinates to the present non-orthogonal coordinates:

$$x^2 = (\varphi - \varphi_p)/ \delta\varphi = 1 + (\varphi - \varphi_s)/ \delta\varphi \tag{3-12}$$

$$\varphi = x^2 \delta\varphi + \varphi_p = (x^2 - 1)\delta\varphi + \varphi_s \tag{3-13}$$

or

$$x^3 - (x^3)_0 = \int_{(x^3)_0}^{x^3} dx^3 = \int_{z_0}^{z} dz / \cos\sigma = l \tag{3-14}$$

where the subscript of 0 denotes the inlet station of computational domain; $z$ is axial coordinate, and $l$ is meridional length; $\sigma$ is meridional flow angle, or is the angle of stream surface on the meridional plane; $\theta_3$ denotes the angle between $x^3$ – coordinate and meridional direction. It and its derivatives are estimated from

$$\left.\begin{aligned}
\tan\theta_3 &= \partial(r\varphi)/\partial x^3 \\
&= x^2\partial(r\delta\varphi)/\partial x^3 + \tan\theta_{3p} \\
&= (x^2-1)\partial(r\delta\varphi)/\partial x^3 + \tan\theta_{3s}
\end{aligned}\right\}.$$

(3-15)

According to the definition of metric tensors we have

$$\sqrt{g_{22}} = \sqrt{((\frac{\partial(r\varphi)}{\partial x^2})^2 + (\frac{\partial}{\partial x^2})^2)} = r\delta\varphi$$

(3-16)

$$\sqrt{g_{33}} = \sqrt{((\frac{\partial(r\varphi)}{\partial x^3})^2 + (\frac{\partial}{\partial x^3})^2)} = \sqrt{\tan^2\theta_3 + 1} = \sec\theta_3 .$$

(3-17)

From above formulas we can see that is a function of $z$ (or $x^3$). Then it is easy to write

$$\partial\sqrt{g_{22}}/\partial x^2 = 0, \qquad\qquad \partial\sqrt{g_{22}}/\partial x^3 = \delta\varphi\sin\sigma + r\partial\varphi/\partial x^3$$

(3-18)

$$\partial\sqrt{g_{33}}/\partial x^2 = \frac{\tan\theta_3}{\sqrt{1+\tan^2\theta_3}}\frac{\partial\tan\theta_3}{\partial x^2}, \qquad \partial\sqrt{g_{33}}/\partial x^3 = \frac{\tan\theta_3}{\sqrt{1+\tan^2\theta_3}}\frac{\partial\tan\theta_3}{\partial x^3}.$$

(3-19)

All the coefficients of the stream function equation can be obtained by substituting equations (3-16) - (3-19) into equations (3-10) and (3-11).

For a given cascade flow field the wall boundary condition is

$$(W)_i = (\frac{\sqrt{g_{33}}}{\rho\sqrt{(g_{11}g_{22})}}\frac{\partial\overline{\psi}}{\partial x^2}G_0)_i . \qquad\qquad (i = p \text{ or } s) .$$

(3-20)

If $(W)_i$ and $\partial\overline{\psi}/\partial x^2$ are known, $\sqrt{g_{33}/g_{22}}$ can be estimated by the boundary condition, Eq.(3-20). As we know $g_{33}$ and $g_{22}$ are interrelated with each other. If the geometry of the pressure side is given, the first step is to calculate the stream function $\overline{\psi}$ everywhere in the flow field of the blade cascade, assuming only a distribution of $g_{22}$ or $\delta\varphi$ along the meridional direction and solving the direct problem. Then, from Eq.(3-20), by giving the target velocity distribution calculate $g_{22}$ or $\delta\varphi$ again. The whole procedure is repeated until convergence of $g_{22}$ or $\delta\varphi$ is reached, which is defined by

$$\left|(\sqrt{g_{22}})^n - (\sqrt{g_{22}})^{n-1}\right| \le \varepsilon_{\sqrt{g_{22}}} .$$

From the calculation procedure it is clearly to see that the stream-function method is based on the imaged plane of $x^1$ and $x^2$, and solving one of the metrical tensor by the boundary condition. In the latter section of potential-function method the same idea is also used.

In [1,2] a set of calculated examples were shown. It demonstrated that this method can be used for compressor and turbine cascade flows. Fig.3-6 Shows a solution of a turbine cascade. Its relative pitch is 0.595. The inlet and outlet flow angles are -44.5 and 57 degrees. The velocity distributions on both pressure and suction surfaces are shown by the dot-solid line. It is seen that the velocity on the suction surface of this cascade profile changes appreciably with. From boundary layer theory we know that the efficiency of this type of velocity distribution can be improved. The modification of the cascade is made by smoothing out the velocity distribution of the suction surface. The new velocity distribution on the suction surface is shown by the dotted line. The resulting profile, calculated by the present method, is shown in the figure by the dotted line. By this method it is enabled to solve the problem of viscous flow[1].

## 4. Stream-Function-Coordinate Method with Target Velocity

The stream-function-coordinate formulation differs from the previous stream-function methods in that the stream functions are taken as coordinates instead of spatial variables. An advantage of this method is extremely convenient for calculating the stream line or stream surface geometry.

In the past few years several papers have been published which give the description of using the Stream-Function-Coordinate (SFC) concept for inviscid flow. Huang and Dulikravich[12] have presented a formulation for 2D and 3D flows. Dulikravich[13] gave the detailed formulation of SFC method for aerodynamic inverse design. The authors[14,15] have presented a generalized method for solving direct, inverse and hybrid methods for turbomachine cascade flows on stream surface of revolution ($S_1$ stream surface flow). It can also be used for tandem cascades or cascades with splitter vanes. For solving 3D aerodynamic problems, Dong and Chen[16] published a paper that gave a new SFC method. In that paper two kinds of stream functions are used. All of the above mentioned papers have shown that the SFC method is an efficient tool for aerodynamic design and analysis of turbomachinery.

### 4.1 Stream-Function-Coordinate Equation

In this section we will discuss an inverse solution method based on 2-D stream-function-coordinate concept. It can be applied to solve both $S_1$ and $S_2$ flows. Here is only $S_1$ stream surface flow of revolution to be concerned. In the following discussion a meridional coordinate system $(\varphi, l)$ is used. $\varphi$ is angular coordinate, and $l$ is meridional coordinate, which is only a function of radius, $r$.

The *continuity* and *momentum* equations of cascade flow on the stream surface of revolution can be written as:

$$\frac{\partial}{\partial \varphi}(\rho \tau W_\varphi) + \frac{\partial}{\partial l}(\rho \tau r W_l) = 0 \tag{4-1}$$

$$\frac{\partial}{\partial l}(W_\varphi r) - \frac{\partial W_l}{\partial \varphi} = -2\Omega r \sin \sigma . \tag{4-2}$$

The definition of rothalpy, equation of state for perfect gas and the formula for calculating density for isentropic process are the same as equations (2-3),(2-4), and (2-5). $W_l$ and $W_\varphi$ denote meridional and circumferential velocity components, respectively;

$\tau$ and $r$ denote stream surface thickness and radius, respectively; $\sigma$ represents meridional angle, $\Omega$ is the angular velocity of rotation.

A dimensionless stream function $\bar{\psi}$ is defined by

$$\left.\begin{aligned} W_\varphi &= -\frac{1}{\rho\tau}\frac{\partial\bar{\psi}}{\partial l}G_0 \\ W_l &= \frac{1}{\rho\tau r}\frac{\partial\bar{\psi}}{\partial\varphi}G_0 \end{aligned}\right\} \tag{4-3}$$

where $G_0$ denotes the flow rate past through a blade channel.

In the flow field of a $S_1$ stream surface of revolution, the stream function is also a function of two independent variables, angular and meridional coordinates. This leads to the coordinate transformation as:

$$\left.\begin{aligned} \frac{\partial(\ )}{\partial\varphi} &= \frac{\partial(\ )}{\partial\varphi}\Big/\frac{\partial\varphi}{\partial\bar{\psi}} \\ \frac{\partial(\ )}{\partial l} &= \frac{\bar{\partial}(\ )}{\partial l} - \left(\frac{\bar{\partial}\varphi}{\partial l}\frac{\partial(\ )}{\partial\bar{\psi}}\right)\Big/\left(\frac{\partial\varphi}{\partial\bar{\psi}}\right) \end{aligned}\right\} \tag{4-4}$$

where: $\bar{\partial}(\ )/\bar{\partial}l$ and $\partial(\ )/\partial\bar{\psi}$ are the derivatives of meridional and circumferential directions along the streamline, which differ from the partial derivatives of $l$ and $\varphi$.

Substituting Eq.(4-4) into Eq.(4-3) we have the following formulas for calculating the velocity components:

$$W_\varphi = \frac{G_0}{\rho\tau}\frac{\bar{\partial}\varphi}{\bar{\partial}l}\Big/\left(\frac{\partial\varphi}{\partial\bar{\psi}}\right) \tag{4-5}$$

$$W_l = \frac{G_0}{\rho\tau r}\Big/\left(\frac{\partial\varphi}{\partial\bar{\psi}}\right). \tag{4-6}$$

From Eqs.(4-4), (4-5) and (4-6) the stream-function-coordinate equation can be obtained:

$$A_1\frac{\partial^2\varphi}{\partial\bar{\psi}^2} + A_2\frac{\bar{\partial}}{\bar{\partial}l}\left(\frac{\partial\varphi}{\partial\bar{\psi}}\right) + A_3\frac{\bar{\partial}^2\varphi}{\bar{\partial}l^2} + A_4\frac{\partial\varphi}{\partial\bar{\psi}} + A_5\frac{\bar{\partial}\varphi}{\bar{\partial}l} = A_6 \tag{4-7}$$

where,

$$A_1 = \left(\frac{\bar{\partial}\varphi}{\partial l}\right)^2 + \frac{1}{r^2}, A_2 = -2\frac{\partial\varphi}{\partial\bar{\psi}}\frac{\bar{\partial}\varphi}{\partial l}, A_3 = \left(\frac{\partial\varphi}{\partial\bar{\psi}}\right)^2$$

$$A_4 = \frac{1}{\rho\tau}\frac{\partial(\rho\tau)}{\partial\bar{\psi}}\left[\left(\frac{\partial\varphi}{\partial l}\right)^2 + \frac{1}{r^2}\right]$$

$$A_5 = \left(\frac{\partial\varphi}{\partial\bar{\psi}}\right)^2\left(\frac{\sin\sigma}{r} - \frac{1}{\rho}\frac{\partial\rho}{\partial l} - \frac{1}{\tau}\frac{\partial\tau}{\partial l}\right)$$

$$A_6 = 2\rho\tau\left(\frac{\partial\varphi}{\partial\bar{\psi}}\right)^3\Omega\sin\sigma/G_0$$

(4-8)

For transonic flow the artificial density technique[18] is used instead of the density in the momentum equation. It can be found by

$$\bar{\rho}_{j,k} = \rho_{j,k} - \mu_{j,k}(\rho_{j,k} - \rho_{j,k-1})$$

(4-9)

where the artificial density sensor $\mu$ is

$$\mu_{j,k} = Max[0, (1 - c/M^2)]$$

(4-10)

where $c$ is the user-specified constant. Typically, $0.8 \le c \le 1.0$. In the transonic case the main difficulty is to avoid the ambiguity in the density when the stream function method is used. For this purpose the meridional velocity components can be calculated from the continuity equation:

$$W_l = \left(\tau W_l \rho r \frac{\partial\varphi}{\partial\bar{\psi}}\right)_{l-1}\bigg/\left(\tau\rho r \frac{\partial\varphi}{\partial\bar{\psi}}\right)$$

(4-11)

where $\rho$ is taken from the last iteration. After $W_l$ is computed, $W_\varphi$ and $W$ can also be obtained.

### 4.2 Boundary Conditions
*Inlet condition*: The inlet flow angle, $\beta_i$, and the upstream velocity $W_i$ are given for all three kinds of problems as

$$\cdot\left(\frac{\bar{\partial}\varphi}{\partial l}\right)_i = \left(\frac{\tan\beta}{r}\right)_i$$

(4-12)

*Outlet condition*: The outlet flow angle, $\beta_e$, is usually given for inverse problem and specified by the Kutta-Zhukovskii condition for both direct and hybrid problems. As in the case of the inlet condition, the outlet condition is also derived, i.e.

$$\left(\frac{\bar{\partial}\varphi}{\partial l}\right)_e = \left(\frac{\tan\beta}{r}\right)_e \cdot$$

(4-13)

*Blade surface condition:* For an arbitrary target velocity distribution a solution exists when the circulation constraint is satisfied. This is an important feature for inverse solutions. The blade surface-condition can be given as

$$\left(\frac{\partial \varphi}{\partial \overline{\psi}}\right)_{wall} = \left[\frac{G_0}{\rho \tau W}\sqrt{\frac{1}{r^2}+\left(\frac{\overline{\partial}\varphi}{\partial l}\right)^2}\right]_{wall}. \qquad (4\text{-}14)$$

The above equation is only for the target velocity distribution. For target thickness distribution a wall boundary condition can be written as

$$\varphi_p = \varphi_s - \delta\varphi \qquad (4\text{-}15a)$$

or
$$\varphi_s = \varphi_p + \delta\varphi \qquad (4\text{-}15b)$$

where $\delta\varphi$ represents the circumferential angular width which stays constant along the meridional direction in the upstream and downstream region.

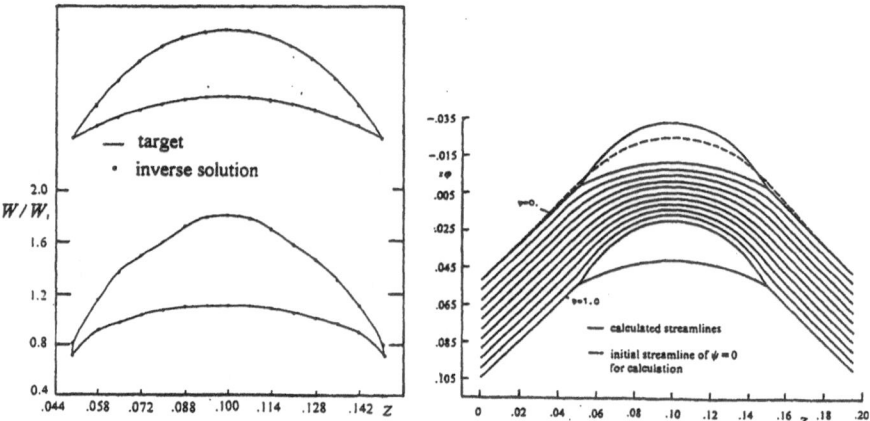

Fig.4-1. Comparison between inverse solution and the target data given by Hobson

Fig.4-2. Stream lines of Hobson's turbine cascade obtained from inverse solution

## 4.3 Calculation Example

The method has been used to calculate a number of blade cascades[14,15,17]. There was published a set of solutions of inverse problems, including a turbine cascade, a compressor cascade, a tandem blade and a mixed-flow impeller with splitter vanes are presented. In the present paper a turbine cascade and a compressor cascade are shown below.

*Turbine blade cascade:* An impulse transonic turbine cascade, which was designed by Hobson, was used to validate the present method. The inlet and outlet angles were $\beta_1 = -46.12$ deg, and $\beta_2 = 46.12$ deg, respectively. The inlet Mach number was 0.575.

The velocity distributions on suction and pressure surfaces are shown by solid curve in Figure 4-1. They were given by Hobson [19], and are taken to be the prescribed quantities as target for inverse design by this method. At the beginning of iterations the initial streamline of $\bar{\psi} = 0$ is given arbitrarily(Fig.4-2). In this case the initial profile thickness is assumed to be zero. The calculated profile by inverse solution coincides perfectly with Honson's profile (the upper figure of Fig.4-1).

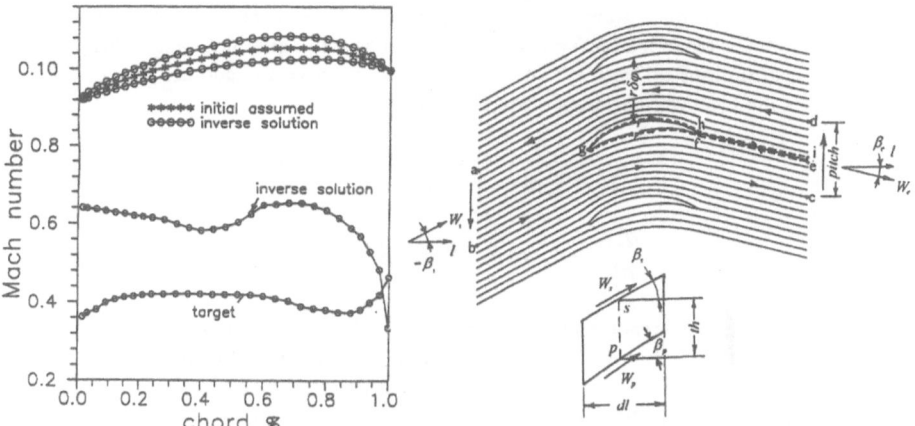

Fig.4-3. Mach number distributions on the blade surfaces and the blade profile.

Fig.5-1. Principle scheme for calculating the blade circulation

*Compressor cascade:* The second example is a conventional compressor cascade. The relative pitch, pitch-to-chord ratio, is 0.657. The inlet and outlet flow angles are -30 deg and -1.15 deg, respectively. The inlet Mach number is 0.509. The angular velocity is 1204.28 sec⁻¹. The Mach number distributions on suction and pressure surfaces are shown in Figure 4-3. The initial profile is taken to be a zero thickness curve.

## 5. Stream-Function-Coordinate Method with Target Circulation

In the previous section the design procedure of blade cascade with stream surface of revolution is based on the prescribed target velocity distributions on two blade surfaces(suction and pressure surfaces) or on one surface. If the target velocity distribution is only on one of these two surfaces, it is necessary to have a thickness distribution. This is need to satisfy the blade strength or the internal cooling requirements. At the preliminary design stage it is more convenient to give the target circulation distribution(loading parameter) than the target velocity distributions since it results to satisfy the blade loading and the turning angle required.

### 5.1 Circulation and Its Derivative of blade

Blade circulation $\Gamma$ is a loading parameter. It characterizes the work done to fluid(or from it) in the blade channel. In Fig. a blade cascade flow field is shown. Fig.5-1 is a principle scheme for derivation of blade circulation and its derivative. The total circulation along a blade(a-b-c-d-a, Fig.5-1) can be written as:

$$\Gamma_0 = -(Wr \sin\beta)_i \delta\varphi + (Wr \sin\beta)_e \delta\varphi$$
$$+ (\Omega r^2 \sin\sigma)_i \delta\varphi - (\Omega r^2 \sin\sigma)_e \delta\varphi \tag{5-1}$$

where subscript $i$ and $e$ denote the inlet of blade channel and the exit from it; $W$ is flow relative velocity; $\Omega$ is angular velocity; $\beta$ is flow angle and $\sigma$ is meridional angle of stream surface; $r$ is radius and $\delta\varphi$ is angular pitch width of one cascade channel which varies along the meridional direction.

It is shown that for a given geometry and rotational speed there exists a unique total circulation(or blade loading) corresponding to a specific inlet flow angle. From the velocity distributions on both surfaces of the blade we have the another formulas for calculating the total circulation of blade(also see Fig.5-1):

$$\Gamma_0 = \sum_{l.e.}^{t.e.} d\Gamma \approx \int_{l.e.}^{t.e.} \frac{d\Gamma}{dl} dl \tag{5-2}$$

$$\frac{d\Gamma}{dl} = -W_s \cos\beta_s + W_p \cos\beta_p \tag{5-3}$$

where $\Gamma$ denotes the local circulation from leading edge to the station; and $d\Gamma / dl$ denotes the first derivative of the local circulation; The subscripts of $l.e.$ and $t.e.$ represent leading and trailing edges, respectively; $s, p$ denote suction and pressure surfaces of blade, respectively.

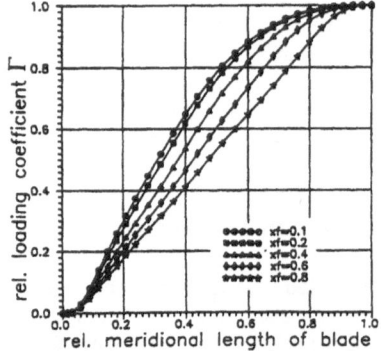 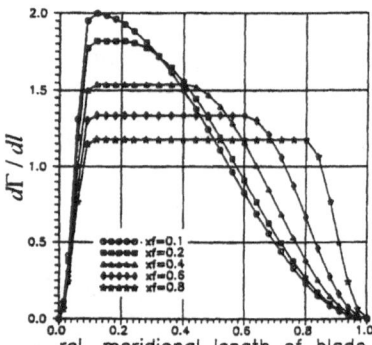

Fig.5-2. Relative blade circulations(or loading coef-   Fig.5-3. Different distributions of first deri-
ficients for different xf.                              vatives of the blade circulation.

From above-mentioned equations for a given total circulation(or blade loading) there are many choices of the distributions of the blade local circulation or its derivative for a compressor cascade(see Fig.5-2, Fig.5-3). At the beginning of design stage it is convenient to give the distribution of the blade circulation or its derivative since the total blade loading is specified. Then, in the present method the blade circulation $\Gamma$ or its derivative $d\Gamma / dl$ can be served as one of the target function. A set of different distributions is shown for demonstration. In Fig.5-2 $xf$ denotes the portion of blade meridional length, after that $d\Gamma / dl$ begins decreasing. For example, $xf = 0.8$, it means that from $l = 0.1$ to $0.8$ $d\Gamma / dl$ keeps constant, and after $l = 0.8$ it decreases to zero at the trailing edge. Then, the greater the value of $xf$, the greater is the portion of the meridional length of blade on where the blade loading is loaded.

## 5.2 Thickness Distribution

The another target function, we choose here, is the thickness of the blade. It is need to satisfy the structure stress condition and cooling(for high temperature turbine) requirements. According the principle of fluid dynamics, if the boundary layer separation does not occur, the flow angles on the surfaces are equal to the geometrical angles of the surface. Therefore, the relationship between the surface angles and the thickness $th$ is:

$$\frac{dth}{dl} = r(\frac{d\varphi_s}{dl} - \frac{d\varphi_p}{dl}) = \tan\beta_s - \tan\beta_p . \tag{5-4}$$

The geometrical relation is also available:

$$\varphi_p = \varphi_s + \delta\varphi - th / r . \tag{5-5}$$

Together with each iterative calculation of the stream-function-coordinate equation both target distributions are enforced to be solved.

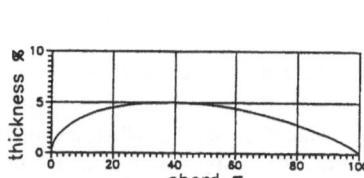

Fig.5-4. Thickness distribution of the compressor blade with 10% relative thickness

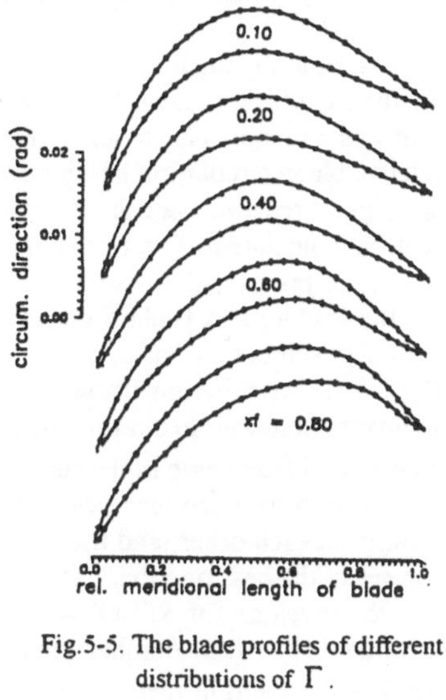

Fig.5-5. The blade profiles of different distributions of $\Gamma$.

## 5.3 Calculation Procedure

The iteration procedure is:
1. to give the geometry, the total circulation, mass flow rate and the inlet aerothermodynamic parameters;
2. to calculate the outlet flow angle by the global mass continuity condition;

3. to give the assumed blade profile configuration;
4. to solve the stream function equation;
5. to calculate the new velocity distributions on both surfaces by Eq.(5-1) and new coordinates of the one surface from another surface by Eq.(5-5);
6. to repeat the steps of 3-5 until convergence being reached.

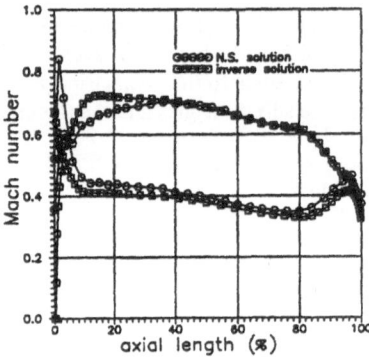
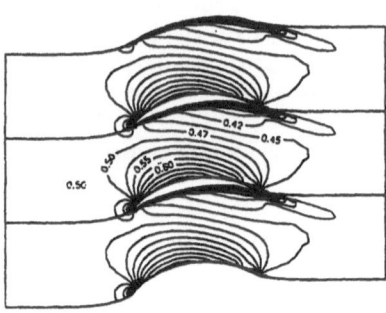

Fig.5-6. Comparison of the Mach number distributions calculated by present and N.S.solution methods

Fig.5-7. Mach number contours calculated by N.S. solver.

### 5.4 Calculation Example

By the use of the present SFC method a set of blade cascades was calculated. One of them was a compressor cascade. Fig.5-2 and Fig.5-3 demonstrates the variations of the relative blade circulation and its derivative from leading edge to trailing edge. One of the target circulation and its derivative is given as a target function, and the other one can be calculated. The target thickness distribution is also given and shown in Fig.5-4. Here the boundary conditions of zero value of the first derivative of circulation at the leading and trailing edges are need to satisfy the inlet flow angle and the Kutta-Jhukovskii condition.

Fig.5-5 shows different blade profiles(with enlarged circumferential size) obtained in the inverse solution. From this figure it is seen that the greater is $xf$, the greater is the change of blade angle in the rear part of the blade channel. Therefore, although the total circulation for each blade cascade is the same, the configuration obtained may be different from each other, and the efficiency is also different. Using N.S. solver or test we can select the best one that has higher efficiency and better off-design performance. One of the versions for xf=0.8 was calculated by the present method. In Fig.5-6 is demonstrated the Mach number distributions on suction and pressure surfaces by the present inverse solution method and also by the N.S. solver[24,25]. Fig.5-7 is the Mach number contours calculated by the N.S. solution method.

## 6. 3-D FLOW FORMULATION OF SFC INVERSE METHOD

In the sections 4 and 5 we have discussed the SFC inverse method for two-dimensional flow. Now, we will extend this approach in three-dimensional flow. A steady inviscid relative flow in a turbomachine can be described by the governing equations, equations (2-1) - (2-10). In this case two kinds of stream functions are applied to describe the flow in turbomachinery.

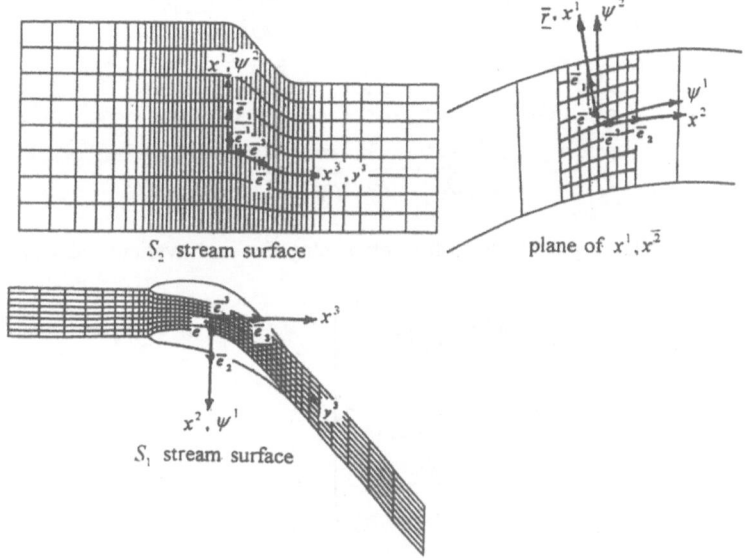

Fig.6-1. Coordinate system used for the 3-D flow solution with two kinds of stream functions and stream-function-coordinate method.

## 6.1 Definition of Stream Functions and Coordinate Transformation

Fig.6-1 shows a typical steam turbine stator and its computational grid system. The stream-function-coordinates are formed in the iteration procedure. In Fig.6-2 are shown three kind of surfaces of this turbine stator. They are: 1)the projection of $S_2$ stream surface coordinate system of $\psi^2$, $x^3$ on meridional plane(upper left); 2)the projection of $S_1$ stream surface coordinate system of $\psi^2$, $x^3$ on the one of blade-to-blade surface of revolution; 3)the meridional plane(in enlarged scale) with $x^1$- and $x^3$- coordinates, and $x^2$-coordinate having the circumferential direction, is perpendicular to $x^1$ and $x^3$.

In order to satisfy the continuity equation (2-6), introduce the following definition for two kinds of stream functions $\psi^1$ and $\psi^2$

$$\rho \vec{W} = -\left(\nabla \psi^1 \times \nabla \psi^2\right) G_0 \tag{6-1}$$

or

$$\rho w^i = \left(\frac{\partial \psi^1}{\partial x^j}\frac{\partial \psi^2}{\partial x^k} - \frac{\partial \psi^1}{\partial x^k}\frac{\partial \psi^2}{\partial x^j}\right) G_0 \tag{6-2}$$

and the relations of the coordinate transformation:

$$\frac{\partial(\ )}{\partial x^1} = -\frac{1}{J}\left(\frac{\overline{\partial}(\ )}{\overline{\partial}\psi^1}\frac{\overline{\partial}x^2}{\overline{\partial}\psi^2} - \frac{\overline{\partial}(\ )}{\overline{\partial}\psi^2}\frac{\overline{\partial}x^2}{\overline{\partial}\psi^1}\right); \tag{6-3a}$$

$$\frac{\partial(\ )}{\partial x^2} = -\frac{1}{J}\left(\frac{\overline{\partial}(\ )}{\overline{\partial}\psi^1}\frac{\overline{\partial}x^1}{\overline{\partial}\psi^2} - \frac{\overline{\partial}(\ )}{\overline{\partial}\psi^2}\frac{\overline{\partial}x^1}{\overline{\partial}\psi^1}\right); \tag{6-3b}$$

$$\frac{\partial(\ )}{\partial x^3} = \frac{\overline{\partial}(\ )}{\overline{\partial}x^3} - \frac{1}{J}\frac{\overline{\partial}(\ )}{\overline{\partial}\psi^1}\left(\frac{\overline{\partial}x^2}{\overline{\partial}x^3}\frac{\overline{\partial}x^1}{\overline{\partial}\psi^2} - \frac{\overline{\partial}x^1}{\overline{\partial}x^3}\frac{\overline{\partial}x^2}{\overline{\partial}\psi^2}\right) + \frac{1}{J}\frac{\overline{\partial}(\ )}{\overline{\partial}\psi^2}\left(\frac{\overline{\partial}x^2}{\overline{\partial}x^3}\frac{\overline{\partial}x^1}{\overline{\partial}\psi^1} - \frac{\overline{\partial}x^1}{\overline{\partial}x^3}\frac{\overline{\partial}x^2}{\overline{\partial}\psi^1}\right) \tag{6-4c}$$

$$J = \frac{\overline{\partial}x^1}{\overline{\partial}\psi^2}\frac{\overline{\partial}x^2}{\overline{\partial}\psi^1} - \frac{\overline{\partial}x^1}{\overline{\partial}\psi^1}\frac{\overline{\partial}x^2}{\overline{\partial}\psi^2}, \tag{6-5}$$

we have the following expressions for determining the contravariant velocity components:

$$\left.\begin{aligned}
w^1 &= \frac{G_0}{J\rho\sqrt{g}}\frac{\overline{\partial}x^1}{\overline{\partial}x^3}\\[2mm]
w^2 &= \frac{G_0}{J\rho\sqrt{g}}\frac{\overline{\partial}x^2}{\overline{\partial}x^3}\\[2mm]
w^3 &= \frac{G_0}{J\rho\sqrt{g}}
\end{aligned}\right\} \tag{6-6}$$

where $x^1, x^2, x^3$ represent curvilinear coordinates. $J$ is the Jacobian determinant of the geometric transformation. $\overline{\partial}(\ )/\overline{\partial}x^3$ is first derivative of $(\ )$ along the streamline on the $x^1, x^2$ surface, which differs from the partial derivative, $\partial(\ )/\partial x^3$.

### 6.2 Stream-Function-Coordinate Equations

Substituting Eq. (6-6) into Eq. (2-7) and Eq. (2-8), two second-order partial differential equations, the stream-function-coordinate equations, can be obtained as follows:

$\overline{e}^1$-direction

$$A_{11}\frac{\overline{\partial}^2 x^1}{\overline{\partial}(\psi^2)^2} + A_{12}\frac{\overline{\partial}^2 x^1}{\overline{\partial}\psi^2\overline{\partial}x^3} + A_{13}\frac{\overline{\partial}^2 x^1}{\overline{\partial}(x^3)^2} + A_{14}\frac{\overline{\partial}x^1}{\overline{\partial}\psi^2} + A_{15}\frac{\overline{\partial}x^1}{\overline{\partial}x^3} + A_{16} = 0 \tag{6-7}$$

$\overline{e}^2$-direction

$$A_{21}\frac{\overline{\partial}^2 x^2}{\overline{\partial}(\psi^1)^2} + A_{22}\frac{\overline{\partial}^2 x^2}{\overline{\partial}\psi^1\overline{\partial}x^3} + A_{23}\frac{\overline{\partial}^2 x^2}{\overline{\partial}(\psi^1)^2} + A_{24}\frac{\overline{\partial}x^2}{\overline{\partial}\psi^1} + A_{25}\frac{\overline{\partial}x^2}{\overline{\partial}x^3} + A_{26} = 0 . \tag{6-8}$$

In the process of solving Eq. (6-7) and Eq. (6-8), the values of both $\psi^1$ and $\psi^2$ vary from 0 to 1. $A_{11}, A_{12}, \cdots$ and $A_{21}, A_{22}, \cdots$ are the coefficients of differential equations. The first subscript denotes component of momentum equation.

The equations (6-7) and (6-8) are used for solving $S_2$ and $S_1$ stream surfaces alternatively. The contravariant components of velocity can be obtained by using Eq. (6-6). Combining equations (2-3), (2-4) and (2-5) the density of each iteration can be estimated by the following formula:

$$\rho = \left[ \frac{B(\rho_{old})^2 - C}{A} \right]^{1/(k+1)}$$
(6-9)

where,

$$A = c_p \frac{(A)^{k-1}}{T_1},$$

$$B = I + \frac{(\Omega r)^2}{2},$$

$$C = \frac{1}{2} \left( \frac{G_0}{\sqrt{gJ}} \right)^2 \left[ g_{11} \left( \frac{\partial x^2}{\partial x^3} \right)^2 + g_{22} \left( \frac{\partial x^2}{\partial x^3} \right)^2 + g_{33} + 2g_{13} \frac{\partial x^1}{\partial x^3} \right].$$

## 6.3 Boundary Conditions

The boundary conditions of the differential equations are as follows:

*Inlet condition:*

The total pressure, the total temperature, the mass flow rate and the distribution of inlet angles are specified.

*Wall condition:*

The stream function $\psi^1$ on the pressure surface is zero and equals unity on the suction surface. The stream function $\psi^2$ is zero at hub and 1 at tip.

*Outlet condition:*

At outlet, the momentum equation is applied. If the outlet gas flow on the meridional plane has an axial direction, the equation can be simplified as:

$$\frac{1}{\rho} \frac{\partial p}{\partial r} = -\frac{V_\varphi^2}{r}$$
(6-10)

where $V_\varphi$ represents the circumferential component of absolute velocity in the cylindrical coordinate system. According to the following relations:

$$V_\varphi = W_\varphi + \Omega r$$
$$W_\varphi = W_z \tan \beta_e .$$
(6-11)

The equation for solving the outlet angle $\beta_e$ can be obtained:

$$\tan \beta_e = \frac{1}{W_z} \left( \sqrt{-\frac{r \partial p}{\rho \partial r}} - \Omega r \right)$$
(6-12)

where subscripts $z$ and $\varphi$ denote axial and circumferential directions, respectively.

## 6.4 Inverse Solution

There are several versions of the target functions given in the inverse solutions: (1) velocity distributions on both surfaces are given; (2) blade thickness distribution and velocity distribution on one of the surfaces are given; (3) blade thickness distribution and pressure difference distribution along the circumferential direction are given; (4) velocity distribution on the tip or hub and main $S_2$ stream surface flow are given, etc.

For the first one, the velocity distribution on the suction and pressure surfaces are specified beforehand. The target velocity distributions on the suction surface and on the pressure surface are represented by $W_{s\,targ\,et}$ and $W_{p\,targ\,et}$, respectively, which can be expressed by

$$W_{i\,targ\,et} = \left[ \frac{G_0}{\rho\sqrt{g}J} \sqrt{\left(\frac{\bar{\partial}x^1}{\bar{\partial}x^3}\right)^2 g_{11} + \left(\frac{\bar{\partial}x^2}{\bar{\partial}x^3}\right)^2 g_{22} + g_{33} + 2g_{13}\left(\frac{\bar{\partial}x^1}{\bar{\partial}x^3}\right)} \right]_i , (i = s\ or\ p) \qquad (6\text{-}13)$$

where subscripts $s$ and $p$ denote the values on the suction surface and pressure surface, respectively, while $g, g_{11}, g_{22}, g_{33}, g_{13}$ denote metrical tensors.

From Eq. (6-13) the following equation determining the shape of the blade surfaces is obtained:

$$\left(\frac{\partial x^2}{\partial \psi^1}\right)_i = \left\{ \frac{G_0}{\rho\sqrt{g}W_i^*} \left[ \left(\frac{\bar{\partial}x^1}{\bar{\partial}x^3}\right)^2 g_{11} + \left(\frac{\bar{\partial}x^2}{\bar{\partial}x^3}\right)^2 g_{22} + g_{33} \right. \right.$$
$$\left. \left. + 2g_{13}\left(\frac{\bar{\partial}x^1}{\bar{\partial}x^3}\right) + \left(\frac{\partial x^1}{\partial \psi^1}\right)\left(\frac{\partial x^2}{\partial \psi^2}\right) \right]^{1/2} \middle/ \left(\frac{\partial x^1}{\partial \psi^2}\right) \right\}_i , (i = s\ or\ p). \qquad (6\text{-}14)$$

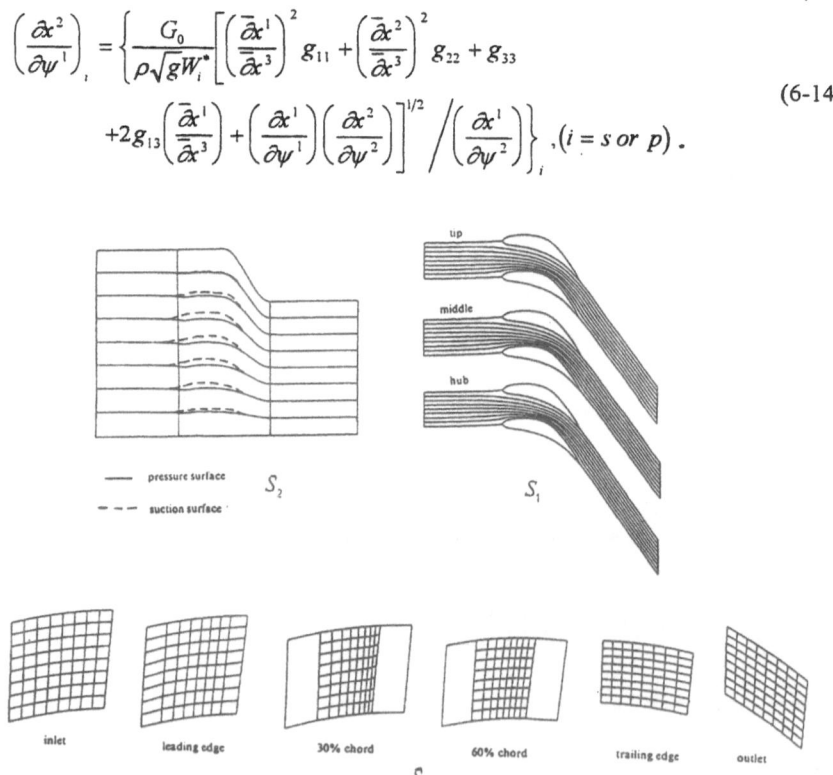

Fig.6-2. Stream lines calculated by the 3-D stream-function-coordinate method.

134

### 6.5 Calculation examples

The 3D flow calculation example is a stator blade row, which consists of 54 blades. It was tested in the annular cascade wind tunnel of HFT, China. The computational grid consisted of $51 \times 9 \times 9$ grid points. The experimental pressure distributions on the suction and pressure surfaces near hub, tip and mid-span are plotted by the squares in Figure 16. The calculated results of the direct solution of this method are agreed with the experimental data well.

For the inverse solution, the calculated velocity from the direct solution is applied to be the target, and the initial geometry of both suction and pressure surfaces are taken to be the same surface(zero thickness assumption). After five hundred iterations the calculation was converged. The calculated suction and pressure surfaces near the hub, at mid-span and near the tip closely reasonably the tested blade. The streamlines of $S_1, S_2$ stream surfaces and plane with coordinates of $x^1, x^2$ are shown in Fig.6-1.

## 7. POTENTIAL FUNCTION METHOD - 3D FLOW FORMULATION

The potential function(PF) method is also based on the principle of imaged plane. The metrical tensors are the variables to be calculated. The difference from the SF method is the use of potential function in stead of stream function. In the present method the potential equation is applied as a main equation to be solved. A special curvilinear coordinate system is adopted for expressing the formulas. As indicated in the previous section, for three-dimensional flow two stream functions are need to be employed. Here only one potential function is applied. This is because of that the continuity equation of three-dimensional flow can be satisfied by only one potential function. In the present PF method the potential function equation is solved by an approximate factorization scheme. It was firstly suggested by Ballaus et al [21]. Zhang [22] used his own improved method for calculating three-dimensional transonic flow in turbomachinery [23]. Then, Dong [16] used this direct method to solve the inverse method. Here we need to point out that for this inverse algorithm the special body-fitted coordinate system is important.

### 7.1 Potential Function Equation

In the present method it is assumed that the absolute flow in turbomachinery is isentropic and irrotational. Then the absolute velocity vector can be written in the form as a function of potential function:

$$\vec{V} = \nabla \Phi = g^{ii} \frac{\partial \Phi}{\partial x^i} \vec{e}_i \quad (i = 1,2,3) . \tag{7-1}$$

The absolute velocity vector can be also decomposed by the contravariant components as

$$\vec{V} = v^i \vec{e}_i \quad (i = 1,2,3) . \tag{7-2}$$

Then the contravariant velocity components are:

$$v^i = g^{ji} \frac{\partial \Phi}{\partial x^i} \quad (i = 1,2,3) \ . \tag{7-3}$$

Substituting above equations into the continuity equation (2-6) the potential function equation is written as

$$\frac{\partial}{\partial x^j}(\rho\sqrt{g}g^{ji}\frac{\partial \Phi}{\partial x^i}) - \frac{\partial}{\partial x^2}(\sqrt{g}\rho\Omega r / \sqrt{g^{22}}) = 0 \ . \tag{7-4}$$

All the equations (7-1)-(7-4) are calculated according to Einstein's summation convention.

In the present method the similar artificial density technique[18] as Eq.(4-9) for three-dimensional flow is introduced by the following form:

$$\bar{\rho} = \rho - \mu(\frac{|W^i|}{W}\frac{\bar{\partial}\rho}{\partial x^i}\Delta x^i) \tag{7-5}$$

where $\frac{\bar{\partial}}{\partial x^i}$ denotes the derivative calculated by the following regulation: when $W^i \rangle 0$, the backward difference is selected, i.e. $\frac{\bar{\partial}}{\partial x^i}$; and when $W^i \langle 0$, the forward difference is used, i.e. $\frac{\bar{\partial}}{\partial x^i}$. The artificial density sensor $\mu$ is defined by the equation (4-10). Finally, the discretized difference potential function equation is written as

$$\left. \begin{aligned} L\Phi_{i,j,k} = \bar{\delta}_{x^1}(\sqrt{g}\bar{\rho}\sum_{i=1}^{3}g^{1i}\delta_{x^i}\Phi)_{i+\frac{1}{2}.j.k} + \\ \bar{\delta}_{x^2}\left[ \sqrt{g}\bar{\rho}(\sum_{i=1}^{3}g^{2i}\delta_{x^i}\Phi - \Omega r / \sqrt{g_{22}}) \right]_{i.j+\frac{1}{2}.k} + \\ \bar{\delta}_{x^3}(\sqrt{g}\bar{\rho}\sum_{i=1}^{3}g^{3i}\delta_{x^i}\Phi)_{i.j.k+\frac{1}{2}} \end{aligned} \right\} \tag{7-6}$$

where the subscripts $i, j$, and $k$ are the grid points along $x^1, x^2$ and $x^3$, respectively. $\bar{\delta}_{x^i}$ is backward-difference operator, and $\delta_{x^i}$ is central-difference operator. This potential function equation is easily to be solved.

### 7-2 Solution Procedure

A special body-fitted non-orthogonal coordinate system, as in the section 3, is chosen and its stream-like coordinate lines are formed by keeping $(\varphi - \varphi_s) / \delta\varphi = const$, where $\delta\varphi$ is angular width of the blade channel; $\varphi_s$ denotes angular coordinate of the suction surface. Then we have:

$$\frac{x^2 - x_s^2}{\Delta x^2} = \frac{\varphi - \varphi_s}{\delta \varphi} \qquad (7\text{-}7)$$

where $x_s^2$ represents the $x^2$ – coordinate(to be zero) on the suction surface; $\Delta x^2$ represents the difference between the $x^2$ – coordinates on pressure and suction surfaces at the same axial station.

In this coordinate system the following partial derivatives can be calculated as

$$\frac{\partial z}{\partial x^2} = 0, \quad \frac{\partial \varphi}{\partial x^2} = \frac{\delta \varphi}{\Delta x^2}, \quad \frac{\partial r}{\partial x^2} = 0 \ . \qquad (7\text{-}8)$$

The metrical tensors can be obtained as:

$$g_{11} = (\frac{\partial r}{\partial x^1})^2 + (r\frac{\partial \varphi}{\partial x^1})^2 + (\frac{\partial z}{\partial x^1})^2, \quad g_{12} = r\frac{\partial \varphi}{\partial x^1}(r\frac{\delta \varphi}{\Delta x^2}),$$

$$g_{13} = g_{31} = (\frac{\partial r}{\partial x^1})(\frac{\partial r}{\partial x^3}) + (r\frac{\partial \varphi}{\partial x^1})(r\frac{\partial \varphi}{\partial x^3}) + (\frac{\partial z}{\partial x^1})(\frac{\partial z}{\partial x^3}) \ ,$$

$$g_{22} = r\frac{\delta \varphi}{\Delta x^2}, \qquad g_{23} = g_{32} = r\frac{\partial \varphi}{\partial x^3}(r\frac{\delta \varphi}{\Delta x^2}), \qquad (7\text{-}9)$$

$$g_{33} = (\frac{\partial r}{\partial x^3})^2 + (r\frac{\partial \varphi}{\partial x^3})^2 + (\frac{\partial z}{\partial x^3})^2 \ ,$$

$$g = \begin{vmatrix} g_{11} & g_{12} & g_{13} \\ g_{21} & g_{22} & g_{23} \\ g_{31} & g_{32} & g_{33} \end{vmatrix} = \left[ \frac{\partial z}{\partial x^1}\frac{\partial r}{\partial x^3} - \frac{\partial z}{\partial x^3}\frac{\partial r}{\partial x^1} \right](r\frac{\delta \varphi}{\Delta x^2})^2 = C(r\frac{\delta \varphi}{\Delta x^2})^2 \ . \qquad (7\text{-}10)$$

The contravariant metrical tensors $g^{ij}$ can be also estimated from metrical tensors $g_{ij}$. In the above formulas $\partial z / \partial x^1$, $\partial z / \partial x^3$, $\partial r / \partial x^1$ and $\partial r / \partial x^3$ keep constant during iterations in the present coordinate system. Then we can see that all the metrical tensors are only the function of $r\delta \varphi / \Delta x^2$. At the beginning of iteration procedure the original blade shape is given. If the geometry of blade is known, all the metrical tensors can be calculated.

With the same principle as the SF method, using the boundary condition, the circumferential angular width can be estimated by the following relationship:

$$\delta \varphi^n = \left[ \frac{\sqrt{g_{11}}\Delta x^2}{Cr}(A^{11}\frac{\partial \Phi}{\partial x^1}r\frac{\delta \varphi}{\Delta x_2} + A^{12}\frac{\partial \Phi}{\partial x^2} + A^{13}\frac{\partial \Phi}{\partial x^3}r\frac{\delta \varphi}{\Delta x^2}) \right]^{n-1} / W^1_{t\,arg\,et} \qquad (7\text{-}11)$$

where: $A^{11} = g_{33} - g_{23} / (r\frac{\delta \varphi}{\Delta x^2})$, $\quad A^{12} = (g_{23}g_{13} - g_{33}g_{12}) / (r\frac{\delta \varphi}{\Delta x^2})^2$ and

$$A^{13} = (g_{12}g_{23})(r\frac{\delta \varphi}{\partial x^1})^2 - g_{13} \ .$$

The subscript $n - 1$ and $n$ denote the old and new iteration values.

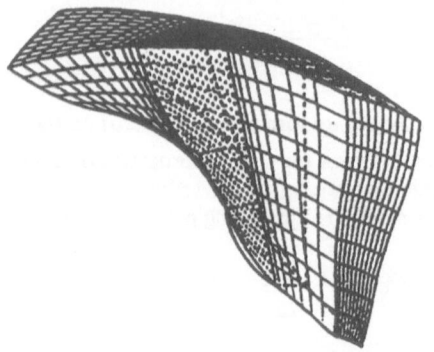

Fig.7-1. A non-orthogonal coordinate system of a compressor rotor.

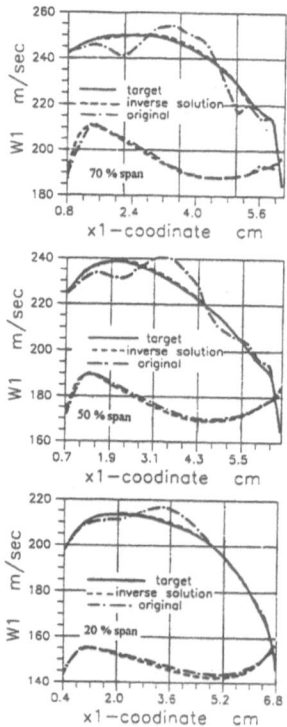

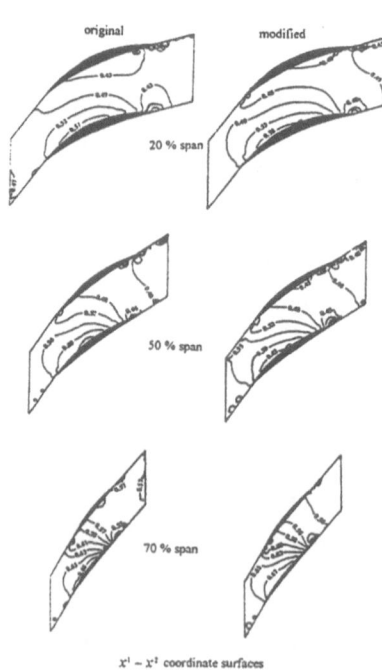

Fig.7-2. Velocity distributions on suction and pressure surfaces of original and modified blades

Fig.7-3. Mach number contours on $x^2, x^3$ surfaces at different span positions

If the $x^1$ – contravariant velocity component of the target velocity $W^1{}_{target}$ is given beforehand, the $\delta\varphi$ can be calculated. Then, the blade geometry can be obtained by successively iterative corrections of the circumferential angular width along the blade span.

## 7.3 Calculated Example

The calculated example is a compressor rotor. It consists of 26 blades. The blade channel is shown in Fig.7-1. The mass flow rate is 40.48 kg/sec at 6699 rpm. The computational grid system of $50 \times 11 \times 11$ nodes is also shown in the figure.

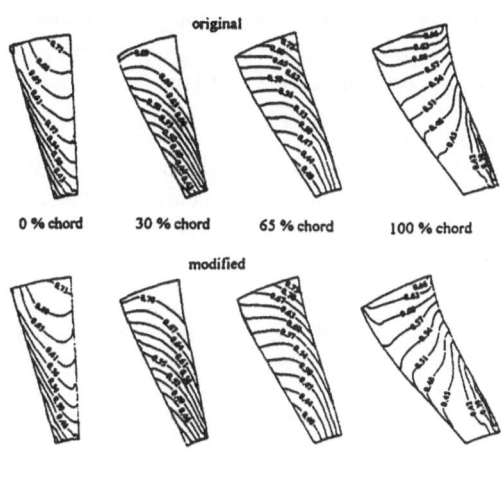

original

0 % chord      30 % chord      65 % chord      100 % chord

modified

$x^2 - x^3$ coordinate surfaces

Fig.7-4. Mach number contours on different $x^2, x^3$ surfaces at
different surfaces of $x^1 = const$

In Fig.7-2 the contravariant velocity components $W^1$ on the suction and pressure surfaces at three different radius(20%, 50%, and 70% of span) are demonstrated. The solid lines are the target velocity distributions(modified). The dotted and solid-dotted curves represent the inverse solution and original velocity distribution. As shown in the figure, the velocity distributions on the blade surfaces after modification are more smooth and very close to the target. Fig.7-3 and Fig.7-4 are the Mach number contours on the different coordinate surfaces. In Fig.7-3 are shown the Mach number contours for the original and modified blades at three span heights. Fig.7-4 is the Mach contours for two blades on the surfaces at different four axial distances of $x^1 = const$.

# 8. INVERSE METHOD BY A DIRECT SOLVER WITH RESIDUAL CORRECTION TECHNIQUE

It is known that in viscous fluid flow there is no velocity at the wall. This is so-called non-slip condition. It makes solution more difficult. Then all the methods concerned with application of stream and potential functions discussed before can not be used in this case. In this section a direct solution method, a Naveir-Stokes solver, is applied for solving inverse method. It is combined with a residual correction. In computation the blade geometry changes in every iteration step. From direct solution the surface pressure(or velocity) distributions differ from the target ones. The difference between these two pressure values at certain station gives a correction of the blade geometry[23]. Then, using the new data of the blade geometry, the computation is carried on again. The iteration procedure is repeated until the solution is converged.

## 8.1 Residual Correction Equation

For two different flow fields in geometry the pressure distributions on the solid walls are different, even though the boundary conditions are the same. The main reason of that perhaps is the differences in coordinates and their derivatives. Then, the correction equation could be written as:

$$\frac{p - p_{t\,\arg et}}{P_{inlet}} = A(y_1 - y_2) + B\partial(y_1 - y_2)/\partial x + C\partial^2(y_1 - y_2)/\partial x^2 + \ldots \quad (8\text{-}1)$$

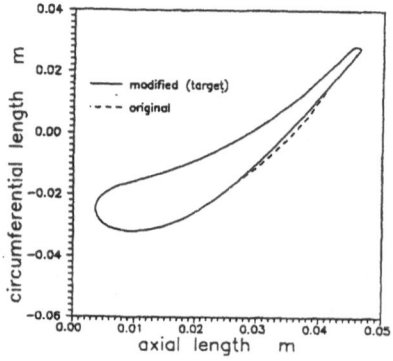

Fig.8-1. A turbine cascade profile

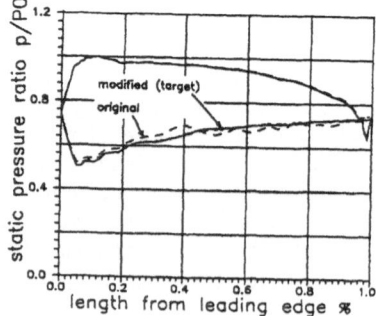

Fig.8-2. Static pressure distributions on the blade surfaces

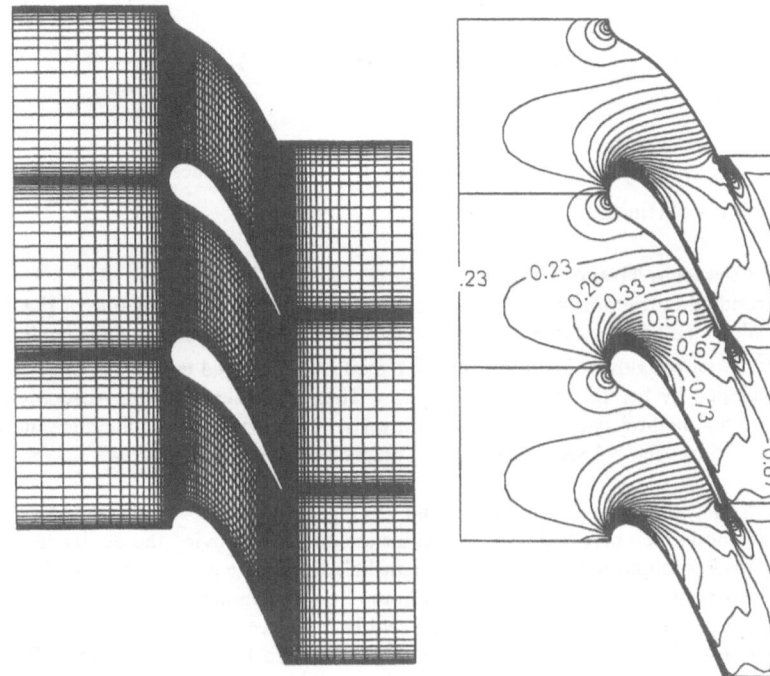

Fig.8-3. Grid system with grid points of 110 × 33.

Fig.8-4. Mach number contours of the modified profile cascade.

In turbomachinery case the pressure difference can be expressed to be a function of one coordinate and its partial derivatives in other two directions:

$$\frac{p_1 - p_2}{P_{inlet}} = A(y_1 - y_2) + B\partial(y_1 - y_2)/\partial x + C\partial^2(y_1 - y_2)/\partial x^2 + ...$$
$$D\partial(y_1 - y_2)/\partial z + E\partial^2(y_1 - y_2)/\partial z^2 + ...$$

(8-2)

Later we will give the residual correction equation only for two-dimensional flow. Neglecting the high-order derivatives with the same reason we have:

$$\frac{p^{n-1} - P_{t\,arget}}{P_{inlet}} = A(r\varphi^{n-1} - r\varphi^n) + B\partial(r\varphi^{n-1} - r\varphi^n)/\partial x + C\partial(r\varphi^{n-1} - r\varphi^n)/\partial x^2 \quad (8-3)$$

where subscripts of $n-1$ and $n$ denote old and new values, respectively; $p_{t\,arget}$ and $P_{inlet}$ represent the target and inlet total pressure, respectively; $A$, $B$ and $C$ denote three empirical coefficients of the differential equation. The dispersed residual correction equation can be obtained:

$$\alpha_{j-1}(r\varphi^{n-1} - r\varphi^n)_{j-1} + \alpha_j(r\varphi^{n-1} - r\varphi^n)_j + \alpha_{j+1}(r\varphi^{n-1} - r\varphi^n)_{j+1} = \frac{p^{n-1} - p_{t \arg et}}{P_{inlet}} \quad (8-4)$$

where $\alpha_{j-1}, \alpha_j$ and $\alpha_{j+1}$ are the relaxation factors. When $p$ reaches to $p_{t \arg et}$, the target geometry is obtained, $|\varphi^n - \varphi^{n-1}| \leq \varepsilon$.

### 8.2 Direct Solution Method

The residual-correction technique can be applied combined with any one of the direct solution methods existed. For viscous flow a time-marching finite volume method[24,25] with Baldwin-Lomax turbulence model is employed in the section. It has been developed by the Institute of Engineering Thermophysics, Chinese Academy of Sciences. This method can be used for both inviscid and viscous calculations, and it was examined by "the NASA Single Rotor Compressor Test Case" in 1994.

### 8.3 Calculation Procedure and Calculated Example

The method given here is demonstrated for the use of improving the old blade cascades. In the computation procedure the original blade cascade is calculated in advance by the direct method, that we selected. From the flow field and the loss calculated we know which portion of the blade profile should be improved. In this case the revised static pressure distributions as the target could be applied to the improvement. From the value of $p - p_{t \arg et}$ and solving the residual-correction equation the displacement at each station can be obtained. Then the calculation by the direct solution for the new geometry is repeated again. This iteration procedure is repeated until solution is converged.

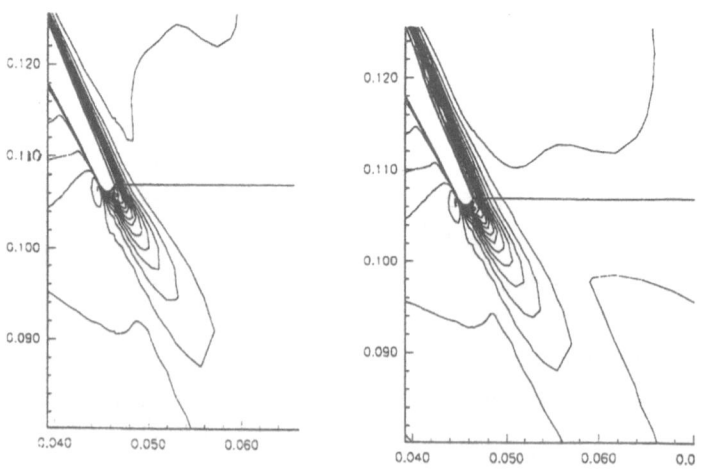

Fig.8-5. Mach number contours of the original blade(left) and
the modified blade(right) in the near-trailing region

The example calculated by the present method is a turbine cascade with round leading edge(see Fig.8-1). The inlet and the outlet isentropic Mach number are 0.3 and about 0.7, respectively. The static pressure distributions on the suction and pressure surfaces are shown in the figure 8-2 by the dotted lines. The pressure distributions of the modified profile, shown also in the figure by the solid lines, are adopted to improve the original profile. The resulted profile by the calculation is given in the figure 8-1 by the solid lines. In Fig.8-4 are plotted the Mach number contours of improved profile cascade obtained by the N.S. solver. Fig.8-5 shows the boundary thickness near to the trailing edge for two blades. It is obviously that the boundary thickness of the original blade is greater than the new one. Then, the loss of the modified blade is smaller than the original blade.

From above discussion we could see that the same idea can be also applied to the three-dimensional flow.

# 9. CONCLUSION

Study in inverse solution problem has great theoretical meaning and practical importance in the improvement of turbomachinery. The methods presented here have been developed in the Institute of Engineering Thermophysics from the last decade. They have been proved to be efficient tools of turbomachinery aerodynamic design. The study of inverse solution can not be developed independently by itself. The method has its own development only when it is tightly connected with the progress of the problems of optimization, fluid mechanics of turbomachinery, computational fluid dynamics, computer science, experimental and measuring techniques, etc. Inverse solution algorithm is being improved with its practical uses.

The author wishes to thank my colleagues, Zhang Fengxian, Xu Yanji, Huang Weiguang, Huang Cunkui, Hai Hao and Li Weihong, from the Institute of Engineering Thermophysics, for their contributions to this paper. The author is also grateful to the National Key Projects on Fundamental Research, China National Natural Science Foundation for the financial supports.

## References

[1] Chen, Naixing, Zhang, Fengxian and Li, Weihong, An Inverse (Design) Problem Solution Method for the Blade Cascade Flow on Stream Surface of Revolution. *ASME* 86-159, or *Journal of Turbomachinery, Trans. ASME*, Vol.108, No.2, pp.194-199, (1986).
[2] Chen, Naixing and Li, Weihong, A Method for Solving Aerodynamic Hybrid Problem of Profile Cascade on $S_1$ Stream Surface of Revolution by Employing Stream-Function Equation Expressed with Non-Orthogonal Coordinate System. *Proc. International Conference on Inverse Design Concepts in Engineering Sciences (ICIDES)*, edited by G.S.Dulikravich, (1984).
[3] Hua, Yaonan and Chen, Naixing, Optimization of the Plane Compressor Blade Aerodynamic Design. *Proc. 6th ISABE*, Paris, (1982).
[4] Wu, Chung-hua, A General Theory of Three-Dimensional Flow in Subsonic or Supersonic Turbomachines of Axial, Radial and Mixed Flow Types, *ASME* Paper no. 50-A-79, *Trans. ASME*, Nov. 1952, or *NACA* TN 2604,(1952).

[5] Wu, Chung-hua, Three-Dimensional Turbomachine Flow Equations Expressed with Respect to Non-orthogonal Curvilinear Coordinates and Methods of Solution, Paper presented at the 3rd *ISABE*, March (1976).

[6] Chen, Naixing, Application of Non-Orthogonal Curvilinear Coordinates to Calculate the Flow in Turbomachinery, *Chinese Journal of Thermophysis*, No.2, (1980).

[7] Liu, Gao-lian and Tao, Cheng, A Universal Image-Plane Method for Inverse and Hybrid Problems of Compressible Cascade Flow on Arbitrary Streamsheet of Revolution, *Power Engineering*(China), No.2, (1981).

[8] Ge, Manchu et al, A Method for Solving Transonic Inverse Cascade Design with a Stream Function Equation, *Journal of Turbomachinery, Trans. ASME*, Vol.108, No.2, pp.200-205, (1986).

[9] Wang, Zhengming, Inverse Design Calculations for Transonic Cascades, *ASME paper 85-GT-6*, (1985).

[10] Luu, T.S., Viney, B., Bencherif, L. and Nguyen Duc, J.M., The Turbomachine Blading Design Using $S_1 - S_2$ Approach, *Proc. International Conference on Inverse Design Concepts in Engineering Sciences (ICIDES)*, edited by G.S.Dulikravich, Oct. 13-15, (1991).

[11] Luu, T.S., Viney, B. and Bencherif, L., Inverse Problem Using $S_1 - S_2$ Approach for the Design of the Turbomachine with Splitter Blades, *Revue Francaise de Mecanique*, No.3, (1992).

[12] Huang, C.Y. and Dulikravich, G.S.,Stream Function and Stream-Function-Coordinate (SFC) Formulation for Inviscid Flow Field Calculation. *Computer Methods in Appl. Mechanics and Eng.*,59,157-177, (1986).

[13] Dulikravich, G.S.,SFC Formulation for Aerodynamic Inverse Design. *AIAA Paper No.91-8019, 29th Aerospace Sciences Meeting*, Reno, Nevada, (1991).

[14] Chen, Naixing And Zhang, Fengxian, A Generalied Numerical Method for Solving Direct, Inverse Hybrid Problems of blade Cascade Flow by Using Streamline-Coordinate Equation. *ASME Paper No. 87-GT-29*, Anaheim, (1987).

[15] Chen, Naixing and Zhang, Fengxian, An Inverse Method for Solving Blade-To-Blade Flows of Turbomachine Tandem Cascades or Cascades with Splitter Vans. *Proc. 4th Asian Congress of Fluid Mechanics*, Hong Kong, (1989).

[16] Dong, Ming, Numerical Solutions of Direct, Inverse and Hybrid Problems of 3D Turomachine Flow. Ph. D. Thesis, Supervised by Professor Chen Naixing, Institute of Engineering Thermophysics, Chinese Academy of Sciences, (1991).

[17] Chen, N.X., Zhang, F.X. and Dong, M., Stream-Function-coordinate(SFC) Method for 2D and 3D Aerodynamic Inverse Design in Turbomachinery, *Inverse Problem in Engineering*, Vol, 1, pp.207-229, (1995).

[18] Hafes, M. and Lovell, D.,Numerical Solution of Transonic Stream Function Equation, *AIAA Journal*, 21,(3) (1983).

[19] Hobson, D.E., Shock-Free Transonic Flow in Turbomachinery Cascades. *Rep. CUED/A Turo/TR40*, University of Cambrige (1972).

[20] Katsains, T. and McNally, W., Fortran Program for Calculating Velocities and Streamlines on a Blade-To-Blade Stream Surface of a Tandem Blade Turbomachine. *NASA TN D-5044* (1969).

[21] Ballaus, W.F., Jameson, . and Albert, J. Implicit Approximate Factorization Scheme for the Efficient Solution of Steady Transonic Flow Problem, *AIAA Journal*, Vol. 18, No.6, pp.573-579, (1978).

[22] Zhang, Jialin, 3D Transonic Potential Flow Computation in An Axial-Flow Compressor Rotor by an Approximate Factorization Scheme, *ASME paper 85-GT-16*, (1985).

[23] Hai, Hao, Numerical Inverse Solution Methods of Two Dimensional Inviscid and Viscous Turbine Cascade Flow, Dissertation of Master Degree, supervised by Professor Chen Naixing, Institute of Engineering Thermophysics, (1996).

[24] Chen, Naixing, Zheng, Xiaoqing and Xu, Yanji, Numerical Computations of Turbomachinery Cascade Turbulent Flows with Shocks by Using Multigrid Scheme, No.*ICAS*-92-3.1.3, (1992).

[25] Chen, Naixing, Zheng, Xiaoqing Huang, Weiguang, Zhang, Fengxian, Zhou, Qian and Xu Yanji, Application of Advanced Numerical Methods to Turbomachinery Aerodynamic Calculations: Time-Marching Method, *Proceeding of the first Asian CFD Conference(* Hong Kong*), Paper No.IF3*, pp. 211-220, (1995).

# Inverse Blade Design Based on Permeable Wall Concept

R.A. Van den Braembussche

von Karman Institute for Fluid Dynamics
Waterloosesteenweg, 72, 1640 Sint-Genesius-Rode, Belgium
e-mail vdb@vki.ac.be

## Summary

Iterative inverse design methods for compressor and turbine blades, with prescribed Mach number or pressure distribution, are presented. It is explained how the permeable wall concept can be used to define the blade modifications required to obtain the target performances with a small number of iterations. It is shown how the method can be combined with potential flow methods, Euler and Navier Stokes solvers. Each method is illustrated with examples. Advantages and problems are discussed and practical solutions to the existancy problem and mechanical constraints are proposed.

## Introduction

Inverse design methods define the blade shape corresponding to a prescribed Mach number or pressure distribution. They are very attractive from the theoretical point of view, as they allow the realization of specified targets, including shock free transonic flows, which can not be realized by the traditional way of try and error. The major drawback is the difficulty to control the mechanical constraints such as minimum/maximum blade thickness, momentum of inertia, etc.

The main problem of inverse design methods is that the boundaries of the numerical domain, in which the flow equations have to be solved, are the solution of the problem. Three techniques are available to overcome this difficulty.

A first one consists of changing variables such that the numerical domain and boundary conditions can be specified as a function of the target. In the hodograph method [1],[2],[3] this is achieved by making the calculations in the q, $\beta$ plane (q is the velocity modulus and $\beta$ is the flow direction). This makes it possible to define the boundaries and boundary conditions of the numerical domain in function of the velocity prescribed on the blade contour.

The numerical domain becomes a simple rectangular when solving the equations in the $\Phi, \Psi$ plane [4],[5],[6]. Boundary conditions define $\Phi$ and $\Psi$ on the walls in function of the target velocity distribution on pressure and suction side:

$$d\Phi = q.ds \qquad \Psi = Cte. \qquad (1)$$

147

After solving the transformed equations, to find q and $\beta$, the blade contour is defined by following integral on pressure and suction side

$$x = \int\limits_{\psi=cst} \frac{\cos\beta}{q}\, d\Phi \qquad y = \int\limits_{\psi=cst} \frac{\sin\beta}{q}\, d\Phi \; . \qquad (2)$$

A second group of methods, called conformal mapping, establishes the relation between a circle, around which the flowfield is known, and the required blade geometry. This relation called mapping function is derived from the relation between the velocity around the circle and the one on the real blade. Methods have been presented by [7] and [8] for single aerofoils and in [9] for turbomachinery blades.

Applications of the methods of these first two groups have been restricted to two-dimensional potential flows because of the important analytical manipulations of the equations that are required. This excludes a lot of practical applications especially in the field of turbomachines where the flow is highly three-dimensional and rotational.

Calculations being performed in a non physical plane (hodograph plane, $\Phi$-$\Psi$ plane or conformal mapping plane), where the blade geometry is not known, makes it difficult to guarantee the geometrical constraints. In some cases it is even difficult to guarantee that the blade is closed and has everywhere a positive thickness [8].

A third group of methods avoiding most of these problems defines the blade directly in the physical plane by defining the numerical domain, and thus also the solution, in an iterative way. They are a logical extention of the traditional try and error method in which the designer decides how to modify the geometry to obtain a target Mach number or pressure distribution. In this group of methods the designers experience and intuition is replaced by a computer algorithm defining the blade modification needed to reach the target. The main challenge is to reach the target as fast as possible.

These methods have known an important development in recent years. They make use of advanced numerical flow solvers (Euler or Navier Stokes) to define iteratively the blade shape and are therefor applicable to a large range of problems. All calculations are done in the physical plane so that the blade geometry is known at each step of the procedure which makes it easier to satisfy the mechanical and aerodynamic constraints.

Numerical optimization schemes have been proposed for this purpose, minimizing the difference between the target pressure distribution and the one calculated on the present geometry [10],[11],[12]. The main advantage of these methods is that they can be combined with any existing flow solver using it as a black box without modification. However these methods have the disadvantage of being very expensive in terms of CPU time because a large number of flow analyses are usually required. Although the use of the adjoint equations [13],[14] allows a considerable gain in computer time the method is still very computer intensive.

During the last decade, several authors developed efficient iterative inverse methods, in which the required geometry modification is defined by means of the flow equations. They

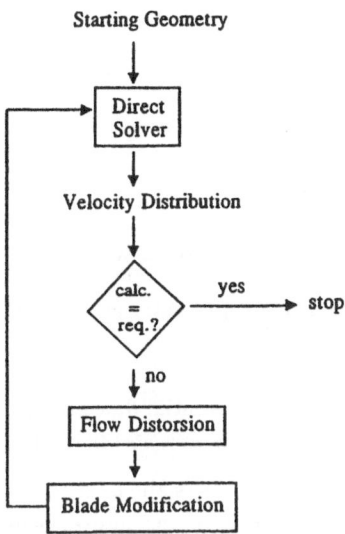

Starting Geometry

Direct
Solver

Velocity Distribution

calc.
=
req.?

yes → stop

no

Flow Distorsion

Blade Modification

Figure 1: Flow chart of inverse design procedure

impose the required pressure distribution on the walls of the present geometry. If this distribution is different from the one resulting from an analysis of the same geometry, the velocity field can not be 'aligned' with the walls. The non zero normal velocity is then used to define a new blade shape by means of a transpiration model. Keeping the walls impermeable but let them move with the flow provides a second way of defining a new blade geometry.

The present paper intends to give a critical overview of some of these blade modification methods and to illustrate them with examples. One will start with the simplest one derived from two-dimensional potential flow equations to illustrate the principle of the method. One will then explain how the same algorithm can be used in combination with an Euler solver and continue with the description of a new three-dimensional method, applicable also to viscous flows.

## Blade modification by means of a singularity method

Following method by Van den Braembussche et al. [15] uses information about the flow, obtained from a singularity method, to define the required geometry modification. Similar methods are described in [16] and [17]. The present method, however, is more general and has been used also in combination with other flow solvers. It consists of following three main components, (Fig. 1) :

- A direct solver to define the tangential velocity $W_t$.

- A flow distortion model to calculate the normal velocity $W_n$ in function of $W_t - W_t^{req}$.

- A blade modification algorithm defining a new blade in function of $W_n$ and $W_t^{req}$.

This sequence of calculations is repeated until

$$W_t = W_t^{req}$$

which means that the present blade geometry produces the required velocity distribution.

**Physical model**

Singularity methods calculate the potential flow around a blade contour by superposing a uniform flow to the one resulting from a vortex distribution on the contour.

The uniform flow is defined in function of the inlet and outlet conditions and the vortices are defined by imposing that the flow is tangent everywhere to the blade contour. It turns out that the velocity on the blade equals the local vortex strength.

$$W_t = \gamma . \tag{3}$$

To change the tangential velocity from its actual value $W_t$ to the required value $W_t^{req}$, one must add a vortex distribution of strength

$$\gamma = W_t - W_t^{req} . \tag{4}$$

In addition to modifying the tangential velocity this additional vortex distribution also creates a velocity normal to the blade contour, which for a cascade of blades at a pitch $t$, is given by :

$$W_n = \frac{1}{2t} \oint \gamma \, H(X,Y) \, ds \tag{5}$$

with

$$H(X,Y) = \frac{\sinh X \cos\beta + \sin Y \sin\beta}{\cosh X - \cos Y} \qquad X = \frac{2\pi}{t}(x - \xi) \qquad Y = \frac{2\pi}{t}(y - \eta) \tag{6}$$

where $\xi, \eta$ are the vortex coordinates and x,y is the location where $W_n$ is calculated.

The discretization of the integral in (5) by a summation over N intervals results in :

$$W_n(i) = \frac{1}{2t} \sum_{j=1}^{N} \gamma(j) \, H(i,j) \, \Delta s(j) \tag{7}$$

and defines the normal velocity in the point "i" due the vortices added in all the points "j".

Streamlines no longer follow the blade and it is shown in next section how to modify the blade shape to align it again with the flow.

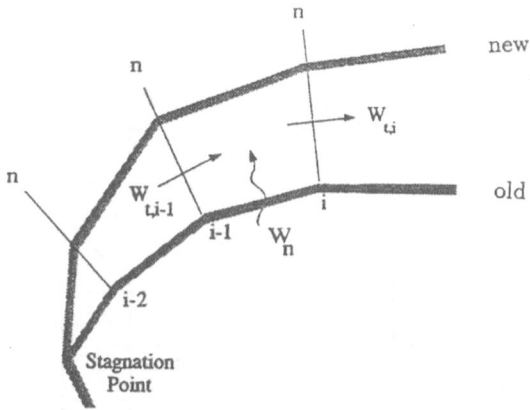

Figure 2: Transpiration model for blade modification

In case these new streamlines are not too far away from the previous blade geometry, the velocity on the new blade will not be very different from the one created by the vortices on the original blade. It can therefor be expected that the velocity on the new blade geometry is closer to the required one than the one of the original blade so that the additional vortices and as a consequence the normal velocity component will become smaller at each iteration.

**Blade modifications**

The geometry modification algorithm calculates the position of the new blade wall in function of the normal and tangential velocity component. This can be done by tracing the wall streamline or by means of a transpiration technique. The last one being more stable, one will describe here only the transpiration technique.

In the transpiration model,the mass balance is applied to the cell defined by the points $(i)^{old}, (i-1)^{old}, (i)^{new}$ and $(i-1)^{new}$ as shown in Figure 2

$$d(\rho \ W_t \ \Delta n) = \rho \ W_n \ ds \ . \tag{8}$$

In discretized form this results in:

$$\Delta n_{i-1} \ \rho \ \hat{W}_t|_{i-1} + \Delta s \ \frac{\rho \ W_n|_i + \rho \ W_n|_{i-1}}{2} = \Delta n_i \ \rho \ \hat{W}_t|_i \ . \tag{9}$$

The ingoing and outgoing velocities $\hat{W}_t|$ are taken as the mean value of the calculated and required tangential velocity along the normal direction respectively at the point $i-1$ and $i$.

$$\hat{W}_t = \frac{W_t + W_t^{req}}{2} \ . \tag{10}$$

Expression (9) allows the calculation of the shift $\Delta n_i$ if $\Delta n_{i-1}$ is known. The modification of pressure and suction side starts at the stagnation point, where the value $\Delta n_1$ is set to

151

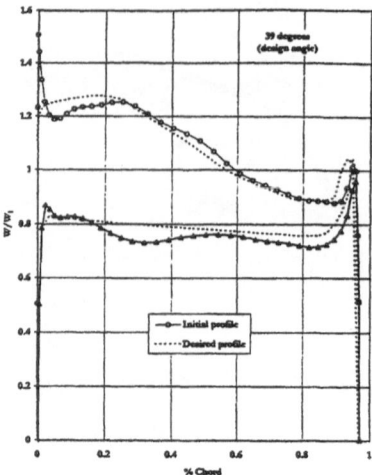

Figure 3: Initial and required velocity distribution of controlled diffusion blade

——— Redesigned Blade

————— Original CDB

Figure 4: Initial and final shape of controlled diffusion blade

zero. As the new blade contour is defined by streamlines one can exclude unphysical blade shapes where the pressure and suction side blade walls cross or do not close at the trailing edge. This could only happen because of errors in the numerical integration procedure or because of inaccuracies in the normal velocity calculation.

## Results

As this method is based on the superposition principle, it is correct only for incompressible potential flows. However experience has shown that it predicts blade corrections which are in the right direction also for a compressible flow, and which anyway vanishe once the velocity distribution has converged to the required one.

This method has been used together with two different direct solvers, a 2D incompressible potential code [18] and a time marching procedure solving the Euler equations [19]. There should also be no problem to combine the present procedure with any existing Navier-Stokes solver as long as the flow is not separated. One example will illustrate the use of each code.

The first example demonstrates the procedure in the design of a compressor blade for

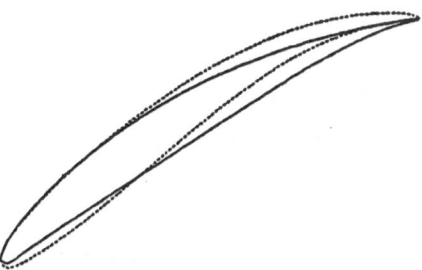

Figure 5: NACA$-65(12A_2I_{8b})10$ (- - -) and NACA$-65(12A_{10})10$ (——) blade shape

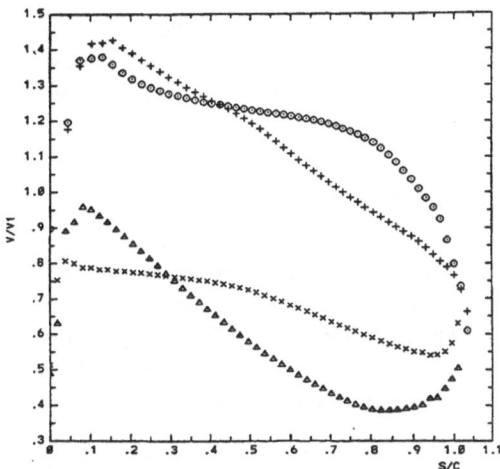

Figure 6: Initial (+,x) and required (o,$\Delta$) velocity distribution

incompressible flow with the required velocity distribution shown in figure 3. The calculations start from an existing controlled diffusion blade [20] at 14.4° stagger (figure 4) of which the calculated velocity distribution is also shown on figure 3

The required blade shape, also shown in figure 4, is obtained after 40 modifications. The CPU time for this example is only a few seconds on a VAX 4000. Although this new geometry is not very different form the initial one (figure 4) the boundary layer calculations and measurements show a considerable decrease of losses [21].

The second example [22] illustrates the combination of the blade modification algorithm, based on singularities, with an existing Euler solver. The new blade is defined by the singularity method and transpiration model, and the compressible flow analysis is made by a time marching procedure applied to an H grid [19].

This combination has been verified by redesigning a NACA-65($12A_2I_{8b}$)10 blade starting from a NACA-65(12A$_{10}$)10 blade as initial guess. The geometries of both blades are compared in figure 5. The velocity distribution on the initial blade and the required velocity distribution are compared on figure 6.

The flow conditions are : $P_1^o = 1.33$ bar, $T_1^o = 341.5°$, $\beta_1 = 45°$, $P_2 = 1.173$ bar. This corresponds to an inlet Mach number of .63 which means that a compressible flow solver is needed to calculate the flow.

Only 12 iterations, composed of one blade modification and one Euler calculation, are needed to obtain an almost perfect agreement between the required and the calculated velocity. The newly designed blade shape is almost identic to the one on which the target velocity has been calculated.

This combination of a singularity method, to define the modification, and an Euler solver for the flow analysis converges to the correct geometry, as long as the singularity method predicts blade corrections that are in the right direction. Experience has shown that this combination can be used up to slightly transonic flows but at the cost of a slower convergence.

Severe convergence problems have been observed when using previous method for the design of transonic blades. Vortices placed at locations where the flow is supersonic, induce a normal velocity also upstream, which is not in agreement with the hyperbolic character of supersonic flow. Difficulties also occur when the flow is rotational, as in radial compressors and turbines where the effect of radius change is more important than the blade curvature. These problems can be avoided by defining also the blade modifications by means of Euler equations, as explained in next section.


## Blade modification by means of an Euler solver


Euler solvers calculate the pressure or tangent velocity distribution on the blade wall, resulting from the boundary condition of zero normal velocity. This velocity field seldomly agrees with the required pressure distribution. Imposing simultaneously the required pressure and zero normal velocity on the blade walls leads to an over-determined problem since one is imposing too many boundary conditions.

The boundary conditions described here reestablishes the well posedness of the problem by allowing a non zero velocity component normal to the blade wall. As already explained the wall needs to be permeable or moving with the fluid. In the present method one will assume permeable walls and use the normal velocity component obtained at full convergence of the time marching calculation for the geometry modification. The same transpiration method, as already explained for the singularity method, will be used for this purpose.

The method has initially been developed for two dimensional flows in cascades [23],[24],[25]. The version presented here is an extension to quasi three dimensional flow on axisymmet-

154

ric stream surfaces with varying radius [26] and applicable to both stators and rotors.

The procedure is almost identic to the one schematically shown on figure 1. The flow distortion module to calculate the normal velocity component, is now an Euler solver with permeable wall boundary conditions.

**Permeable wall boundary conditions**

Each solution of the two-dimensional time-dependent Euler equations can be written as the superposition of 4 waves, propagating in the numerical domain [24]. The propagation velocities in the direction $\vec{n}$, normal to a given boundary, are the eigenvalues of the Jacobian matrix associated to the direction $\vec{n}$ (outward pointing normal vector). The compatibility relations, which are a reformulation of the Euler equations along the wave front, may be used to define the new values of the unknown.

The four eigenvalues and the corresponding compatibility relations are :

| Eigenvalue | Compatibility relation |
|---|---|
| $W_n^{n+1}$ | $\rho^{n+1} = \rho^* + \frac{P^{n+1} - P^*}{a^{n,2}}$ |
| $W_n^{n+1}$ | $W_t^{n+1} = W_t^*$ |
| $W_n^{n+1} + a^n$ | $W_n^{n+1} = W_n^* - \frac{P^{n+1} - P^*}{\rho^n a^n}$ |
| $W_n^{n+1} - a^n$ | $W_n^{n+1} = W_n^* + \frac{P^{n+1} - P^*}{\rho^n a^n}$ |

The value * refers to the value calculated by means of an eccentric version of the numerical scheme of the solver applied to the boundary of the numerical domain. The compatibility relations can be used only if the wave approaches the wall from inside the domain (positive eigenvalue). Negative eigenvalues mean that the wave is entering the domain. As a consequence the corresponding compatibility relation, containing information from inside the domain, can not be used and a boundary condition is needed to define the unknown at the boundary.

The number of boundary conditions therefor equals the number of negative eigenvalues and depends on the sign of $W_n$. The later one can be calculated from the third compatibility relation because the corresponding eigenvalue $W_n + a$ is always positive, (one assumes that the normal velocity component can not be supersonic). The pressure $P^{n+1}$ in the third relation is obvious the required pressure.

If $W_n$ is positive, only one boundary condition (the required static pressure ) must be imposed, since only one eigenvalue ($W_n - a$) is negative. If $W_n$ is negative, two additional boundary conditions are needed, since also the first and second eigenvalue are negative. It has been found that the best solution is to impose the relative total pressure and total temperature at the points where $W_n < 0$.

In real turbomachinery flows the stagnation conditions change from inlet to outlet because of work input and shock losses in transonic flows. One will therefor impose the value of the stagnation pressure and temperature calculated at the previous analysis step.

As the total pressure does not change from one iteration to the next, the new static temperature can be defined from the change in static pressure by means of the isentropic relation.

$$T^{n+1} = T^n \left(\frac{P^{n+1}}{P^n}\right)^{\frac{\gamma-1}{\gamma}}.$$

The velocity can then be defined by

$$W^2 = 2C_p(T^{o,n} - T^{n+1})$$

and the tangential velocity component is given by

$$W_t^2 = W^2 - W_n^2 .$$

Once the normal velocity is known, a new blade shape can be defined by means of the same transpiration method as described before.

## Applications

Any Euler solver can be used to redesign a blade, providing it is accurate and its boundary conditions have been adapted. The one used in following examples is based on a finite volume time marching procedure, using a centered scheme [24],[26]. Both second order and fourth order viscosities are used for the time discretization. A fourth order Runge-Kutta scheme is used for the time integration. A cell vertex approach on non-intersecting finite volumes is used for the space discretization. C-grids are used for a better leading edge modelling in case of turbine blades. H-grids are used for high speed compressor blades with sharp leading edges.

The method solves the quasi three-dimensional Euler equations in quasi linear form on an axisymmetric surface, taking into account the variation of the streamtube height $h(m)$. The boundary conditions are modified to allow the flow to cross the permeable blade walls when required.

The first example demonstrates the procedure in the redesign of a turbine blade at transonic flow conditions. The starting geometry is the VKI-LS 82-05 turbine blade [19] operated at $P_1^o = 1$ bar, $T_1^o = 278.°$, $\beta_1 = 0.°$, $M_2 = 1.0$.

A shockless transonic Mach number distribution has been imposed on the suction side (figure 7). The calculated Mach number distribution converges as close as possible to the required one (figure 8). Complete convergence could not be achieved, probably because there is no physical geometry corresponding to the required Mach number distribution. The problem of existence of a solution is out of the scope of this paper but is discussed in more detail [25] and [28]. The original and final geometries are compared in figure 9. The main difference is the suction side curvature.

Besides several axial compressors and turbines the method has also been used to redesign the blade section near the shroud of a centrifugal compressor, operating at 26000 RPM (figure 10) [27].

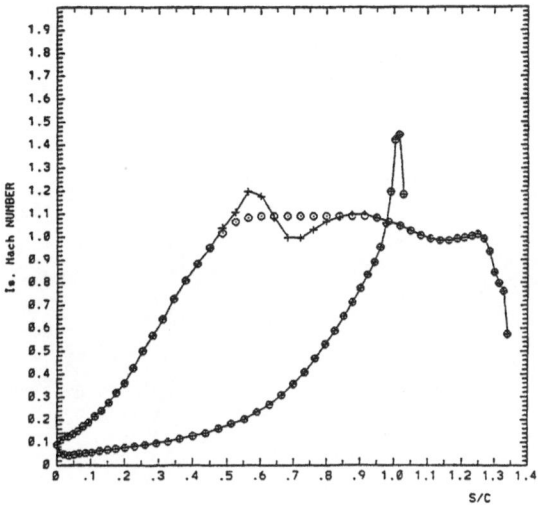

Figure 7: Initial (+) and required (o) Mach number distribution

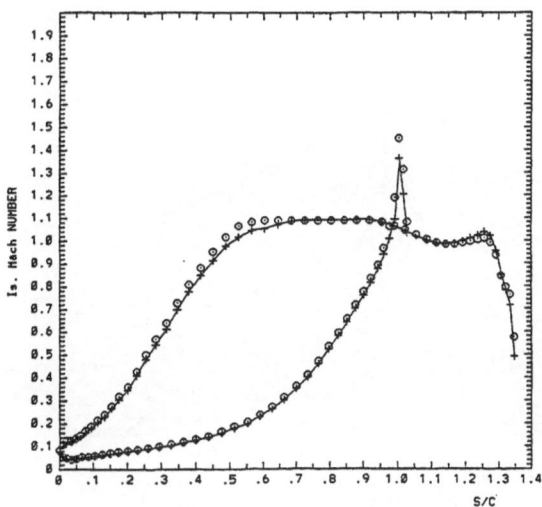

Figure 8: Calculated (+) versus required (o) Mach number distribution after 2 modifications

Figure 9: Original (grey) and redesigned (——) blade shape

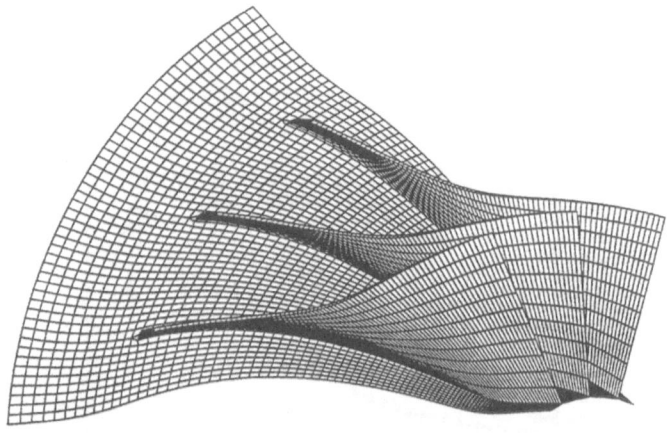

Figure 10: Radial impeller geometry

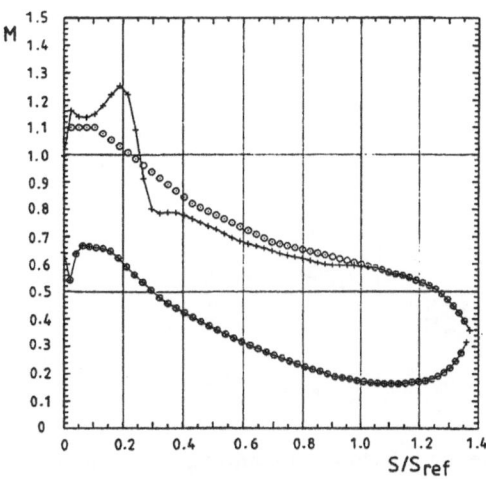

Figure 11: Initial (+) and required (o) Mach number distribution at shroud

Input for the quasi 3D design program is the geometry of the stream surface $(r(z))$, the initial blade suction and pressure side position on the stream surface $(\theta_{ss}, \theta_{ps})$ and a fictitious streamtube height defined by

$$h(m) = \frac{1}{(2\pi \; r - Z \; th/cos\beta_{bl} \; \rho(m) \; \overline{W}(m) \; cos\beta(m)} \; . \tag{11}$$

The H-grid on the 3D flow surface is obtained by the inverse transformation of the H-grid developed around the 2D transform of the shroud blades.

$$dx = \frac{dm}{r} \qquad\qquad dy = \frac{rd\theta}{r} \; . \tag{12}$$

Results of a blade-to-blade analysis, shown on figure 11, indicate a rapid expansion around the leading edge, up to a peak Mach number of 1.25 , followed by a strong deceleration on the blade suction side due to a bow shock. The interaction of the last one with the suction side boundary layer could provoke boundary layer separation. As Coriolis and curvature forces prevent the flow from reattachment downstream, this could have a detrimental effect on the performance and stability of the impeller.

The method has been used to redesign the shroud section of previous impeller, imposing a shock free Mach number distribution with lower maximum Mach number as shown on figure 11. It is expected that the new blade will result in an important improvement of the performance at high pressure ratio.

The iso-density lines obtained by a subsequent analysis show a much weaker bow shock which is no longer touching the suction side boundary layer (figure 12).

The method described in this section can easily be extended to three-dimensions and has

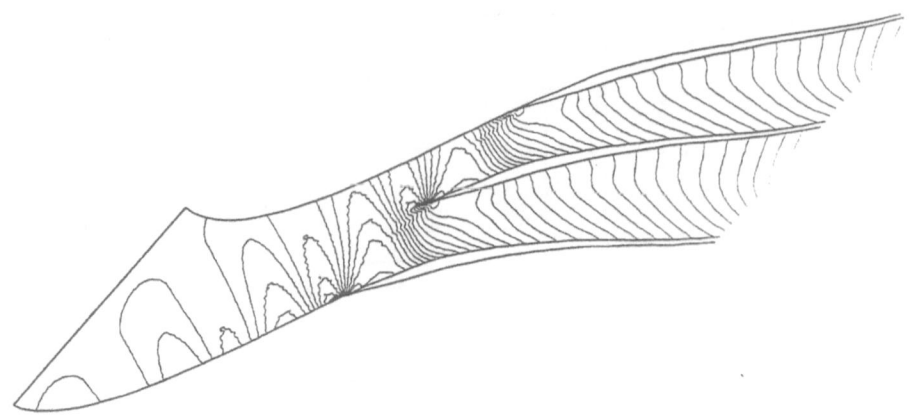

Figure 12: Isodensity lines on the shroud streamsurface for the redesigned blade surface

successfully been used for the design of three-dimensional turbine and compressor blades [29],[30].

## Three dimensional blade design

The main disadvantage of the method described in previous chapter is that it provides a new approximation of the blade geometry only after full convergence of the Euler solver. A non negligible amount of computation is required before one can verify if the constraints are respected. This makes it difficult to update the target pressure distribution when the new geometry is not acceptable. Moreover the permeable wall boundary concept is questionable when applied to viscous flow computations. As a matter of fact, one can hardly imagine to calculate a boundary layer, with zero tangential velocity, on a permeable wall with a non zero normal velocity component. The method has therefore been reformulated and the strategy of calculating the inviscid or viscous flow on an impermeable moving wall has been adopted. One calculates the transient which occurs when the blade walls move to their required location.

Other inverse methods following the 'moving wall' strategy [31],[32] also modify the blade shape each time the wall boundary conditions are applied. However, they can not take into account the additional terms resulting from the mesh movement, because the wall modification is a result of the flow analysis. This does not result in an error because those terms become zero as soon as the inverse method has converged to the required blade shape. However adding those terms results in more accurate flow predictions and permits larger blade shape modifications per iteration. As a consequence the method is more stable and converges faster to the target [33],[34],[35]. For this reason, the geometry modification and flow field updates have been separated in this new formulation (figure 13). Each iteration consists of a blade shape and mesh modification, followed by a time

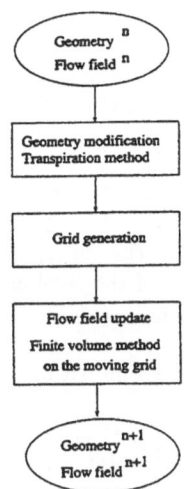

Figure 13: Flow chart of moving wall inverse design method

step of the finite volume solver to update the flow field on the moving mesh and wall.

The finite volume method needs to be somewhat 'generalized' when analysing the flow on a moving mesh because the conservation relations are applied on control volumes which move and whose shape may change from one time step to the other. This induces additional terms in the equations, which are function of the mesh point movement and volume change. They can be calculated from the known mesh position before and after the shape modification.

The wall boundary conditions applied during the analysis step, are the classical solid wall conditions, which are well known and established for viscous as well as for inviscid problems. Except for the wall movement velocity, they are the same along the whole boundary and do not create discontinuities in the wall.

The geometry modification algorithm being separated from the flow field computation, it can be based on any concept. Previous experience has shown that the transpiration model is very efficient to define the required geometry changes, and it is therefore also used in this new method. The blade shape modification is calculated again by transpiration in function of the velocity component normal to the permeable wall.

In this formulation, the target pressure is imposed only during the blade modification step and the blade walls are modified as long as the calculated pressure distribution is different from the prescribed one. The flow analysis converges to a steady state when the blade shape remains unchanged because the prescribed pressure distribution is reached or when further blade shape modifications are prevented by the mechanical constraints.

## The blade shape modification

The blade shape modification starts with the calculation of the normal velocity which is then input into the transpiration method. One will first explain the inviscid version of the method. The particularities related to the viscous version of the method will be exposed next section.

**Permeable wall boundary conditions:** As explained previously each solution of the 3D Euler equations can be written as the superposition of 5 waves propagating in the numerical domain. The 5 eigenvalues of the jacobian matrix associated to the direction $\vec{n}$ and corresponding compatibility relations are:

| Eigenvalue | Compatibility relation |
|---|---|
| $W_n^{n+1}$ | $\rho^{n+1} = \rho^* + \frac{P^{n+1}-P^*}{a^{n,2}}$ |
| $W_n^{n+1}$ | $W_t^{n+1} = W_t^*$ |
| $W_n^{n+1}$ | $W_s^{n+1} = W_s^*$ |
| $W_n^{n+1} + a$ | $W_n^{n+1} = W_n^* - \frac{P^{n+1}-P^*}{\rho^n a^n}$ |
| $W_n^{n+1} - a$ | $W_n^{n+1} = W_n^* + \frac{P^{n+1}-P^*}{\rho^n a^n}$ |

Where $W_n, W_s$ and $W_t$ are respectively the velocity component in the normal, streamwise and spanwise direction (figure 14) and the state $*$ refers again to the extrapolated values.

As already explained for the two dimensional case one can use the compatibility relation to calculate the unknown on the boundary of the numerical domain if the eigenvalue is positive. A third boundary condition is required when $W_n < 0$. .

As for two dimensional flows one can again use the isentropic flow equations to derive the static temperature from the imposed static pressure. The magnitude of the relative velocity, as well as the velocity component tangent to the blade surface are defined in the same way as before.

$$W^2 = 2C_p(T_r^{0,n} - T^{n+1}) \qquad W_s^2 + W_t^2 = W^2 - W_n^2 \qquad (13)$$

where $T_r^o$ is the total temperature in the relative frame of reference.

The additional boundary condition to define the additional unknown, is the ratio between the stream- and spanwise velocity components, which is also imposed at its local value during the previous analysis step:

$$\left(\frac{W_s}{W_t}\right)^{n+1} = \left(\frac{W_s}{W_t}\right)^n. \qquad (14)$$

Those three boundary conditions may not provide an exact solution at each iteration, but will not falsify the final results, as they are updated after each geometry modification. The final result is obtained anyway by an analysis step.

**3D Transpiration model:** The transpiration model defines a new blade geometry in function of the calculated normal and tangential velocity along the initial blade suction and pressure side.

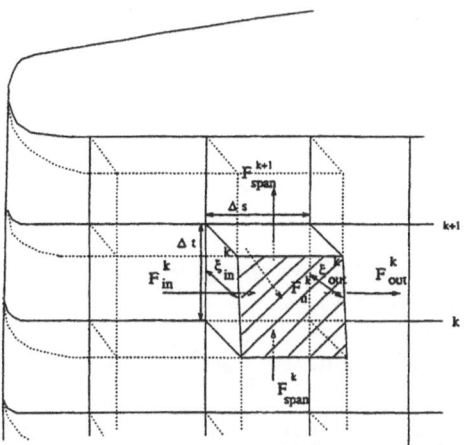

Figure 14: 3D transpiration

The method progresses separately along the pressure and suction surfaces, from the stagnation line to the trailing edge, keeping the stagnation line on the blade unchanged. All mesh points, including those on the blade wall, are displaced in the axisymmetric surfaces on which they have been specified. As these surfaces are defined by a linear interpolation between hub and shroud they remain unchanged during the whole design.

The amplitude of the displacement is obtained by applying the mass equilibrium in the cells between the old and new blade walls (figure 14). The flux through the elementary surface ($\Delta s \Delta t$) of the old blade wall is defined by:

$$F_n^k = (\rho W_n \Delta s \Delta t)^k \tag{15}$$

where $\Delta s$ and $\Delta t$ are the dimensions of the face in the stream- and spanwise directions. Assuming that the new blade wall is a streamsurface, the streamwise fluxes are:

$$F_{in}^k = (\rho W_s \Delta t)_{in}^k \xi_{in}^k \qquad F_{out}^k = (\rho W_s \Delta t)_{out}^k \xi_{out}^k . \tag{16}$$

The spanwise mass fluxes $F_{span}$ through the interior surfaces account for the interaction between the neighbouring cells:

$$F_{span}^k = (\rho W_t \Delta s)^k \frac{\xi_{in}^k + \xi_{out}^k + \xi_{in}^{k-1} + \xi_{out}^{k-1}}{4} . \tag{17}$$

The hub and tip walls being impermeable, no mass flux can enter or leave the two extreme cells through them. The density and velocity components are obtained by interpolation, from the values at the calculation nodes. A three-dimensional grid, containing $KM$ surfaces from hub to tip, results in a tri-diagonal system of $KM-1$ equations, defining the unknowns $\xi_{out}^k$ in function of the displacements $\xi_{in}^k$, the last one being already computed from the mass equilibrium of the upstream cells.

The displacements of the $KM$ grid points are then derived from the displacements $\xi_{out}^k$, by interpolation for the interior points, and by extrapolation for the hub and tip points.

## The flow field update

After the blade shape modification, a new mesh is generated around the new profile. The displacements of all mesh points over the time step are computed, from which a 'grid-point velocity' field is deduced.

The conservation of mass, momentum and energy is applied to control volumes $V$, whose faces are moving with the velocity $\vec{V_g}$:

$$\frac{d}{dt}\int_V \rho dV = \oint_S \rho(\vec{W} - \vec{V_g})\vec{n}dS \tag{18}$$

$$\frac{d}{dt}\int_V \rho W_i dV = \oint_S [\rho W_i(\vec{W} - \vec{V_g})\vec{n} + pn_i]dS + \int_V F_i dV$$

$$\frac{d}{dt}\int_V \rho E_r dV = \oint_S [\rho E_r(\vec{W} - \vec{V_g}) + \vec{W}p]\vec{n}dS$$

where $\vec{n}$ is the inward pointing normal vector.

Splitting the time derivative into two terms, the discretized form of the mass conservation equation (18) becomes:

$$\Delta\rho \frac{V}{\Delta t} = \sum_{faces} \rho\vec{W}\vec{n}\Delta S - \left(\sum_{faces} \rho\vec{V_g}\vec{n}\Delta S + \rho\frac{\Delta V}{\Delta t}\right). \tag{19}$$

The first right-hand-side term is the residual for a fixed mesh, while the two additional terms are due to the mesh movement.

Both $\frac{\Delta V}{\Delta t}$ and $V_g$ are function of the grid movement and not independent. The relation is readily obtained from the Space Conservation Law (SCL) expressing the mass conservation in a flow field with zero velocity [37].

$$\frac{d}{dt}\int_V dV + \oint_S \vec{V_g}\vec{n}dS = 0 . \tag{20}$$

It is clear that if this relation is not satisfied, erroneous mass sources or sinks are introduced, leading to instabilities. Not-satisfying the SCL therefore requires to under-relax the mesh points displacements per iteration, or induces an additional constraint on the time step size, which may be more severe than the temporal discretization stability constraint. A detailed derivation of this relation in a axisymmetric coordinate system is given in [34] and [35].

The two additional terms, resulting from the mesh movement, converge to zero at steady state and it is clear that they should also compensate each other in smooth flows. It could therefor be concluded that the presence of those terms is not necessary for steady computations, as proposed in [31] and [38]. However, experience has shown that these additional terms favor a stable and monotonic convergence, especially in transonic flows with shocks.

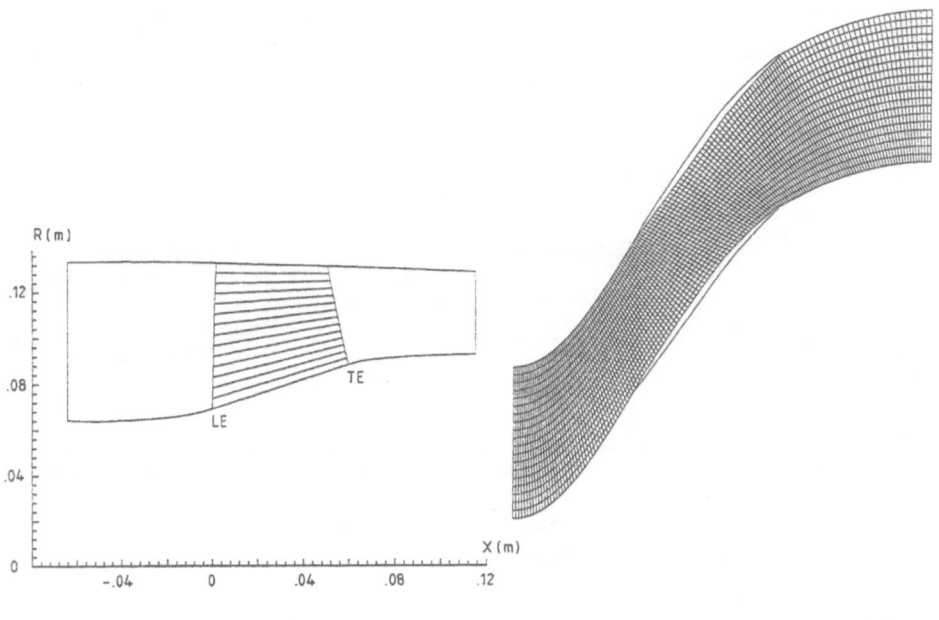

a. Meridional view          b. Non periodic H-grid

Figure 15: Transonic compressor rotor blade - Initial design

## Applications

This new design procedure has been implemented into a high resolution three-dimensional solver, written for this purpose [29],[30],[34] and has been used for the design of axial and radial turbomachinery [36].

Flux vector splitting [39] is used for the computation of the advective terms in the cell centered finite volume approach. The unsteady equations are integrated in time with an explicit four-step Runge-Kutta algorithm. M.U.S.C.L. reconstruction is used to obtain a higher order accuracy.

The three-dimensional structured grid is composed of two-dimensional ones, defined on a series of axisymmetric surfaces between the hub and shroud endwalls. Non reflecting boundary conditions [40] are applied along the inlet and outlet boundaries. The outlet static pressure is specified by the user at one spanwise position between the hub and tip endwalls, and calculated at the other stations, to respect the radial and axial momentum equations.

The design of transonic compressor blades is certainly a domain where inverse methods can bring substantial performance improvements, by allowing the design of shock free blades. The method is first illustrated by the design of the complete transonic axial compressor rotor with 14 twisted blades, rotating at 31264 RPM [33],[35].

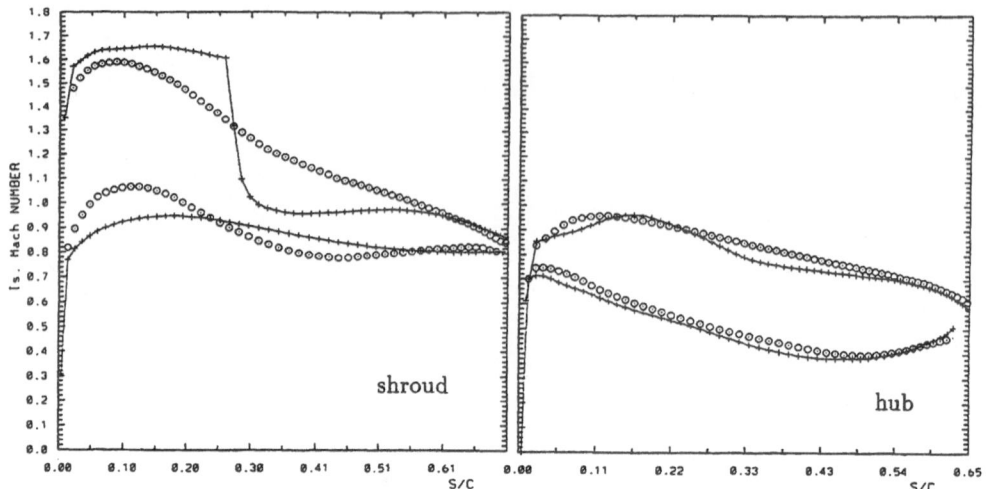

Figure 16: Transonic compressor rotor - Calculated (+) and target (o) Mach number distribution at shroud and hub section

The meridional view (figure 15a) shows a non negligible contraction of the passage, and indicates the 15 axisymmetric surfaces on which non periodic H-grids (figure 15b) have been generated to discretise the numerical domain

The inlet relative Mach number varies from 0.74 at the hub section to 1.35 at the shroud. Isentropic Mach number distributions along the hub, mean and shroud section of the original blades are shown in figure 16.

The blade has been redesigned, to achieve a shock free isentropic Mach number distribution. The last one is specified at five equidistant sections between the hub and shroud, and calculated at all grid points by interpolation between these five sections. The prescribed distributions are compared to the initial ones in figures 16.

The Mach number distributions are optimized to achieve a controlled diffusion with shock-free deceleration on the suction side. The pressure side has a much smaller influence on the performances because of its lower Mach number and limited diffusion. It has therefor not been prescribed but calculated by the solver in order to satisfy the mechanical constraints. Each time a new suction side is defined by the transpiration method, the pressure side is modified simultaneously in order to keep the blade thickness distribution as prescribed. In this way the method itself predicts what the pressure side Mach number is needed to satisfy the constraints. Once the solution approaches convergence (after 500 iterations) the calculation is continued with the normal design procedure whereby the previously calculated Mach number is imposed on the pressure side. An additional 1000 iterations are required to obtain a perfect agreement between the calculated and imposed pressure distributions.

The new prescribed pressure side Mach number distribution shows more diffusion than the

166

initial one, and is therefore less optimum from the aerodynamic point of view. However, imposing an optimum pressure side Mach number as target, with less diffusion, would have resulted in a too thin (thick) front (rear) part of the blade shroud section as a consequence of the lower (higher) average velocity. Satisfying the thickness constraint by adjusting the suction side Mach number distribution should have resulted in a less efficient blade.

The hub, mean and tip cross-sections of the initial and final blade geometries are compared in figure 18. One observes that the blade thickness distribution is conserved at each cross section.

The final iso-density lines at the shroud section are compared to the initial ones in figures 17. One can observe that the same pressure ratio is achieved, with a much smoother compression along the redesigned suction surface. The figure illustrates how the S shaped suction side generates compression waves, converging into a bow shock, which does no longer interact with the suction surface.

Euler solvers do not account for viscous effects, and the suction and pressure side boundary layer displacement thickness has to be subtracted from the designed blade to obtain the geometrical blade. This explains why the initial and final blade shapes shown in figure

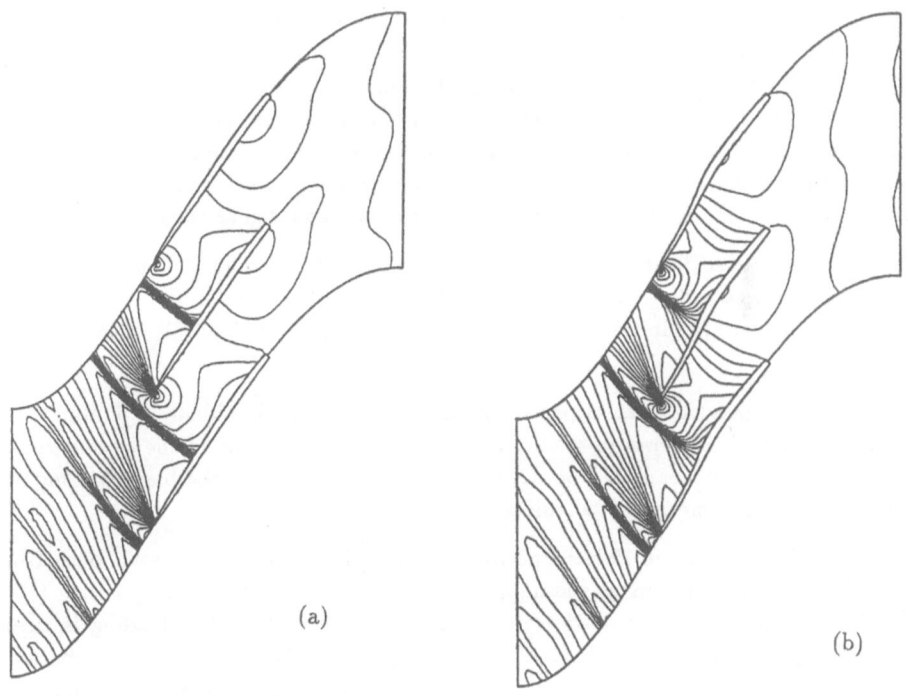

(a)

(b)

Figure 17: Isodensity lines at shroud of the initial (a) and redesigned (b) transonic compressor rotor

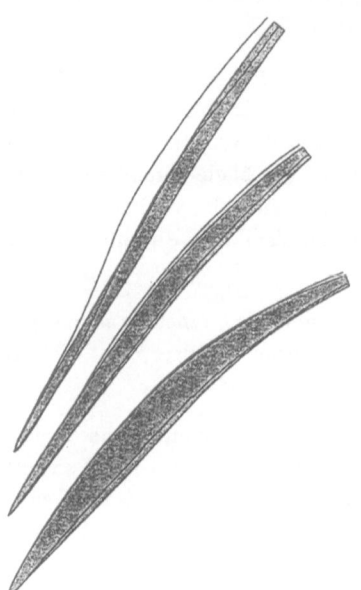

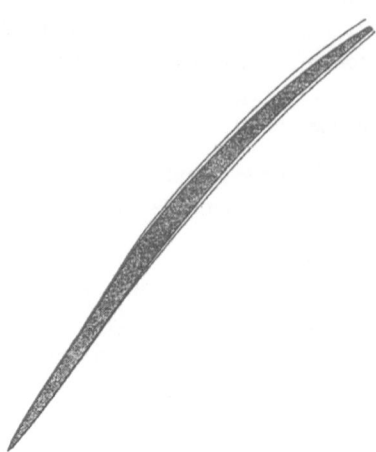

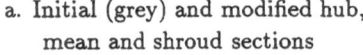

a. Initial (grey) and modified hub,
mean and shroud sections

b. Boundary layer subtraction

Figure 18: Blade geometry modification

18a have thick trailing edges. The trailing edge thickness corresponds to the sum of the geometrical and boundary layer displacement thicknesses, which can already be calculated from the imposed Mach number distribution, before the blade geometry is known. After completing the inverse design of the blade, the boundary layer displacement thickness must be subtracted, to obtain the shape of the metal blade (figure 18b).

The second test case illustrates the use of the 3D inverse method for the optimization of a turbine annular cascade with 21 blades, an aspect ratio of 0.7, and a pitch-to-chord ratio of 0.9 at midspan. The initial geometry has a constant VKI-LS89 profile from hub to tip [41].

The flow through such a low aspect ratio highly loaded turbine blade is dominated by secondary flows. It is shown in the literature [42] that the intensity of the secondary flow can be decreased by means of a positive compound lean, reducing the blade loading near the endwalls, but increasing it at midspan.

The blade shown on figure 19 has a positive compound lean of 30 degrees. The isentropic Mach number distributions at the hub, mean and shroud section, corresponding to the radially stacked and leaned blades are compared on figure 20. The loading is clearly reduced near the endwalls, and increased at midspan. The Mach number distribution is improved at the hub and shroud, but is definitely not optimal at the mean section.

The leaned blade has been redesigned with the 3D inverse solver, to achieve a smooth acceleration along the suction side as shown in figures 21. The three different target

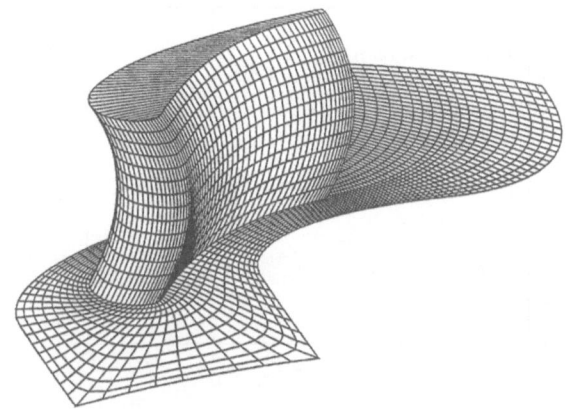

Figure 19: 3D view of the leaned blade

Mach number distributions, specified on the suction side of each section, illustrate an option that allows imposing the outlet flow angle or required trailing edge thickness. The three Mach number distributions are function of a parameter which is modified during the design process until the preset outlet flow angle or trailing edge thickness is obtained.

Following expression shows that for a prescribed mass flow $(Q_m)$ and trailing edge Mach number, the trailing edge thickness and outlet flow angle are related.

$$Q_m = \rho M_{te} a_{te} cos\beta \left( t - \frac{th}{cos\beta} \right) \tag{21}$$

where $t$, $th$ and $\beta$ refer to the pitch, the blade thickness and flow angle. Adjusting the outlet flow angle by changing the free parameter simultaneously fixes the trailing edge thickness.

Full convergence inclusive adjustment of $\beta_2$ is obtained after 4000 iterations only. The initial and final blade shapes are compared at the hub, mean and tip sections in figure 22a, and the initial and final spanwise distributions of pitchwise averaged outlet flow angle are compared in figure 22b.

## Inverse design by means of a Navier Stokes solver

Following section demonstrates the feasibility to combine previous design system with a Navier Stokes solver. The viscous version of the inverse method follows exactly the same strategy as the inviscid one, except that the flow analysis is now made with a Navier Stokes solver instead of Euler solver.

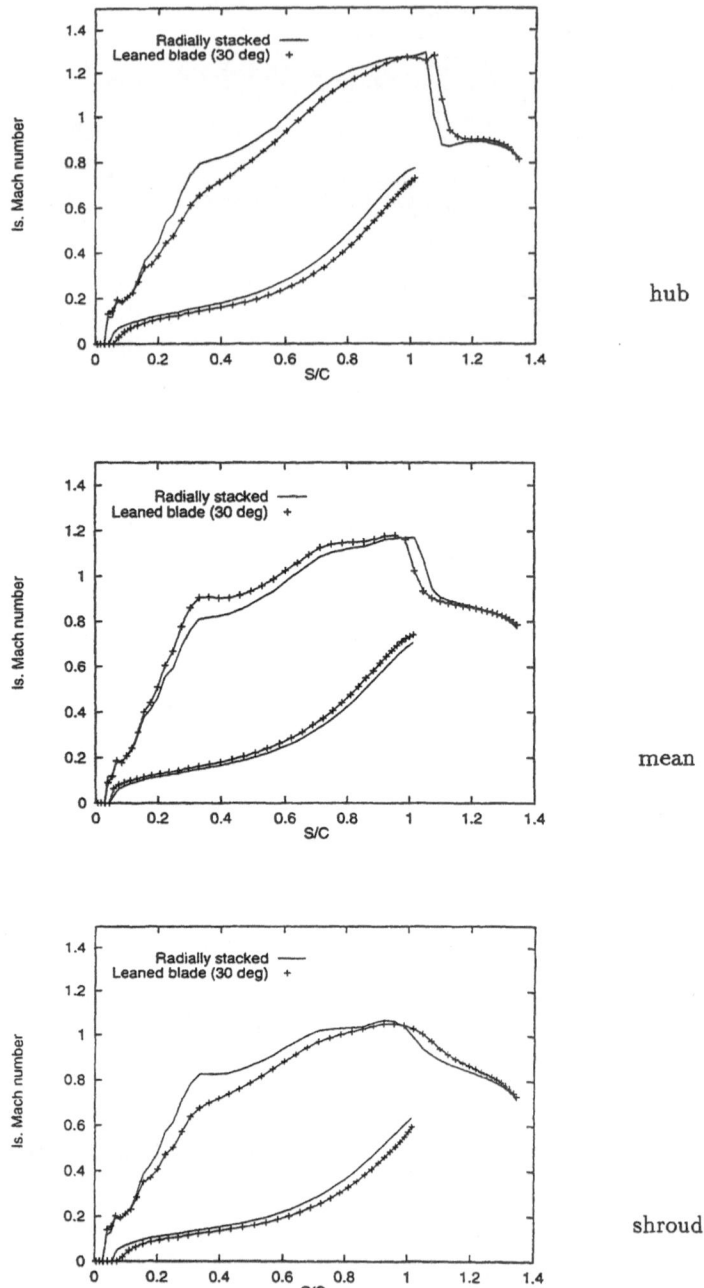

Figure 20: Mach number distribution on radially stacked (——) and leaned (+) blades at hub (a), mean (b) and shroud (c)

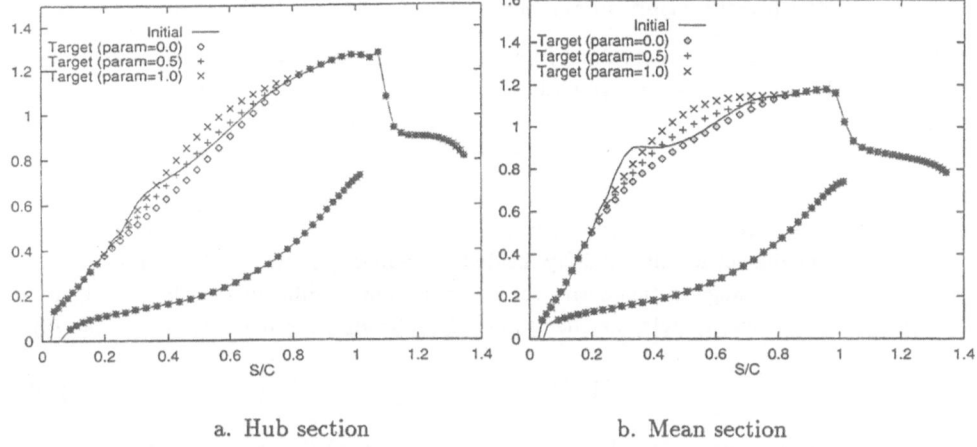

a. Hub section            b. Mean section

Figure 21: Inverse calculation - Initial and prescribed is. Mach number distributions

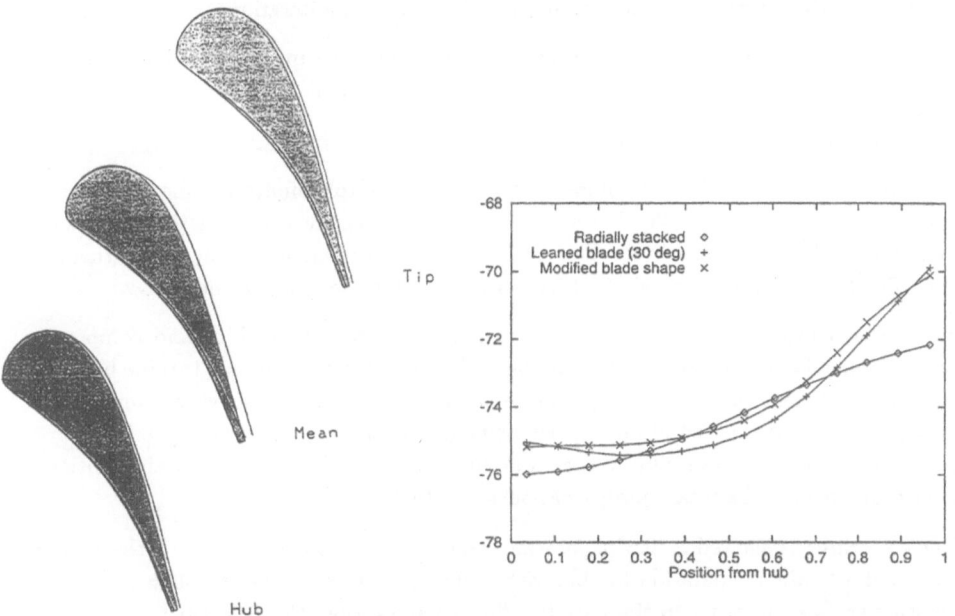

a. Initial (grey) and modified blade shapes       b. Outlet flow angle distribution

Figure 22: Blade shape and outlet flow angle modification

## The geometry modification algorithm

The normal velocity is again obtained from the compatibility relation associated with the eigenvalue $W_n + a$, which is always positive.

$$W_n^{n+1} = W_n^* - \frac{P^{req} - P^*}{\rho^n a^n} \ .$$ (22)

Where the state $^*$ refers again to the value obtained by extrapolation from inside the domain. The change of normal velocity inside the boundary layer is of the same order of magnitude as the change in tangential velocity and is more difficult to calculate than in an Euler solver. It was therefor decided to use the following approximate version of (22):

$$W_n^{n+1} = -\frac{P^{req} - P^{act}}{\rho^n a^n}$$ (23)

relating the normal velocity distribution to the difference between the prescribed pressure and the one calculated by the Navier Stokes solver. This new calculation of the normal velocity has the advantage of not requiring a time-marching iteration.

The tangential velocity component used in the transpiration model is obviously not the one on the wall but the velocity at the outer edge of the boundary layer.

## Applications

Previous procedure has been combined with the Navier-Stokes finite volume solver developed by O. Léonard et al. [43] The space discretization scheme is exactly the same as the one used in the 3D Euler solver. The molecular viscosity is modelled by the Sutherland law. The Baldwin-Lomax algebraic turbulence model is used for turbulent flow.

The method has been successfully used for the design of both turbine and compressor blades [44],[45] and is illustrated here by the redesign of the LA subsonic turbine blade for the flow conditions specified in [46]. The calculations have been made with a multi-block grid (figure 23), composed of an hyperbolic O-grid near the profile, an algebraic H-grid to discretize the wake, an elliptic C-grid around the O- and H-block with an additional H-grid upstream. The total number of nodes is ±16500.

The experiments showed a flow separation near 83% of the axial chord on the suction side, and a laminar separation bubble with turbulent reattachment at about 15% of the pressure surface. In order to simulate this flow configuration, the Baldwin-Lomax model has been activated, respectively at 70% of the axial chord on the suction side and at 10% on the pressure side.

The calculated isentropic Mach number distribution on the blades is compared to the target one in figure 24. The main goals of this prescribed distribution are a reduction of the suction side maximum Mach number, and the suppression of the two velocity peaks observed in the leading edge region. Perfect agreement between the final pressure distribution and the target is reached after a few thousand iterations. The initial and final geometries are shown in figure 25 together with a detailed view of the leading edge.

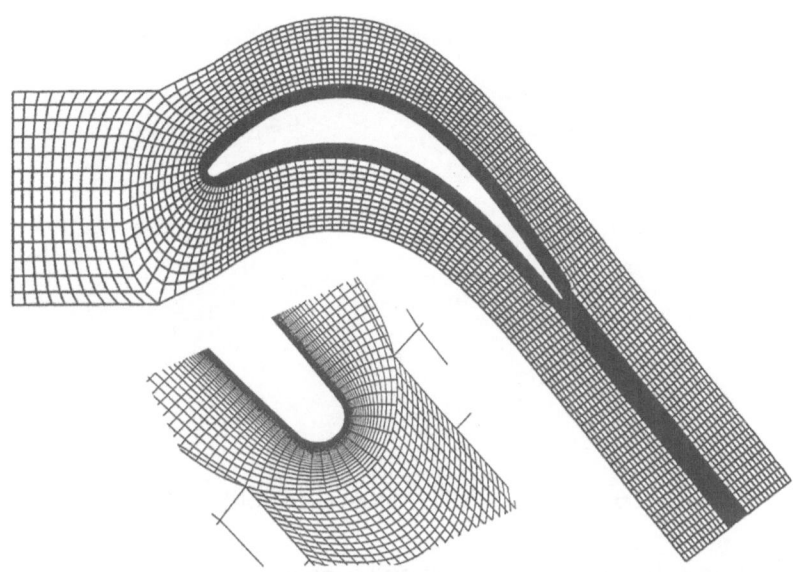

Figure 23: General view and detail of the multi-block mesh around LA turbine blade

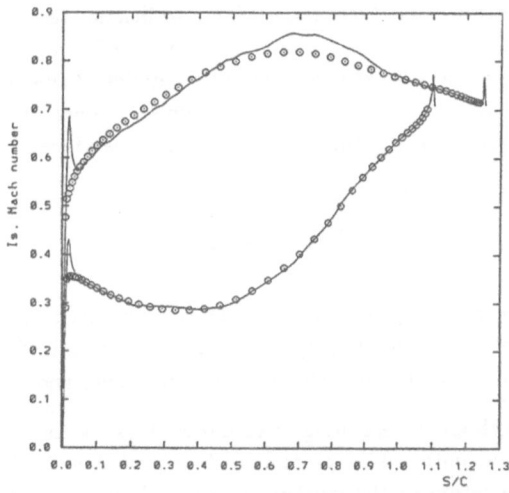

Figure 24: Calculated (——) and imposed (o) Mach number distribution on the LA turbine blade

During the inverse calculation, the outlet isentropic Mach number has been iteratively and automatically modified (final value: 0.8367), so that the mass flow through the cascade remained constant.

As already pointed out, the additional driving terms resulting from the mesh points movement contribute to a better convergence. They are used simultaneously with local

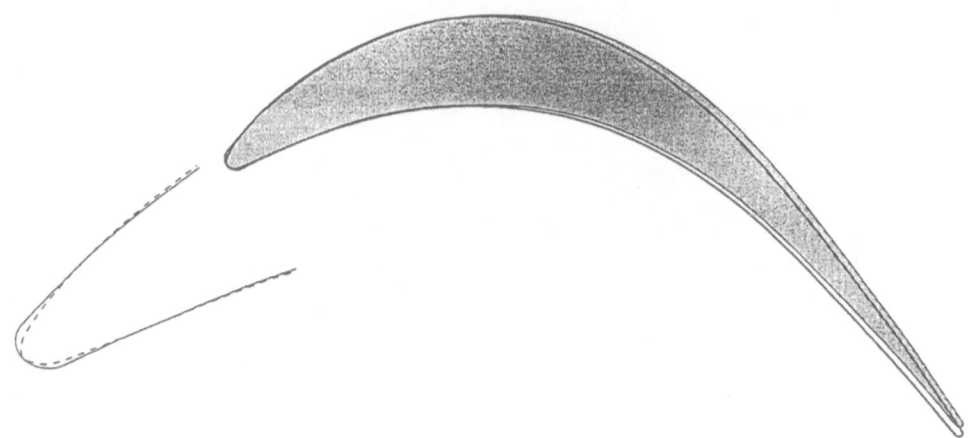

Figure 25: Initial (grey) and final blade shape with zoom of leading edge

time stepping, enthalpy damping (for inviscid computations only), and residual averaging. None of these techniques influence the steady asymptotic solution. However, local time stepping in highly stretched grids, as used for Navier stokes calculations, easily results in inconsistent mesh displacement velocities. The additional terms become unphysical in the wall region, where the time step changes rapidly with the mesh, and perturb the convergence. For this reason, they are only activated outside of the viscous layers, so that they can still be used in combination with local time stepping. Convergence histories show a definite improvement if the additional terms are introduced.

## Conclusions

It has been show how geometry modification algorithms based on the permeable wall concept can be used efficiently for the inverse design of 2D and 3D blade shapes. It is explained how they can be combined with potential flow, Euler and Navier Stokes solvers.

Compared to other methods, where integral values such as loading are prescribed, these methods provide a direct control of the pressure or Mach number distribution on the blade. This allows the optimization of the compressor or turbine performance by controlling directly the diffusion in the boundary layers.

The optimum pressure distribution is well known for two-dimensional flows [9]. Defining it for complex 3D geometries is not an easy task and presently a topic of research. Numerical optimization methods can find such an optimum but they have the disadvantage of requiring much more computational effort.

Making the design in the physical space facilitates the control of the mechanical constraints. The pressure may be prescribed on part of the blade surface only, keeping the rest of the geometry unchanged. Special features allowing the adjustment of the target

Mach number distribution to satisfy the constraints on outlet flow angle and trailing edge thickness have been demonstrated.

The method is applicable to both viscous and inviscid design problems. It shows a rapid convergence to the blade geometry corresponding to the target pressure distribution, for both subsonic and transonic flow. In case such a solution does not exist the method converges to the closest possible solution.

The effects of blade lean are taken into account in the 3D flow calculation, which is an advantage, when compared to quasi three-dimensional methods.

An important advantage of the method is the possibility to use the same solver for both the design and analysis of an optimized blade.

All methods can easily be used also for single aerofoil design [47]

## Acknowledgements

The author wants to acknowledge the important contribution of Dr. O. Léonard and Dr. A. Demeulenaere to the developement of the methods presented in this contribution.

## References

[1] Bauer F., Garabedian, P., and Korn, D.: Supercritical Wing Sections, Vol. I, Springer-Verlag, New York, (1972).

[2] Cantrell, H.N. and Fowler, J.E.: The aerodynamic design of two dimensional turbine cascades for incompressible flows with high speed computer, ASME paper 58-A-141, (1958).

[3] Sanz J.M.: Automated design of controlled diffusion blades, ASME Journal of Turbomachinery, Vol. 110, No 4, pp 540-544, (1988).

[4] Stanitz, J.D.: Design of two dimensional channels with prescribed velocity distributions along the channel walls, NACA TR 1115, (1953).

[5] Schmidt, E.: Computation of supercritical compressor and turbine cascades with a design method for transonic flows, ASME Journal of Engineering for Power, Vol. 102, pp 68-74, (1980).

[6] Varonos, A., Chaviaropoulos, P. and Papailiou K.: A design method for stator cascades with streamsurfaces of revolution using natural coordinates, Inverse Problems in Engineering, Vol. 2, pp.119-139, (1995).

[7] Lighthill, J.M.: A new method of two dimensional aerodynamic design, ARC R&M 2112, (1945).

[8] Volpe, G.: Transonic shock free wing design, Inverse Methods for Airfoil Design for Aeronautical and Turbomachinery Applications, AGARD R 780, Paper 5, (1990).

[9] Papailiou, K.D.: Blade optimization based on boundary layer concepts, von Karman Institute CN 60, (1967).

[10] Vanderplaats, G.N.: Approximation concepts for numerical airfoils optimization, NASA TP-1370, (1979).

[11] Eyi, S. and Lee, D.: High-lift design optimization using the Navier-Stokes equations, AIAA paper 95-0477, (1995).

[12] Thibert, J.J.: One point and multi-point design optimization for airplane and helicopter application, Inverse methods for Airfoil Design for Aeronautical and Turbomachinery Applications, AGARD-

R-780, (1990).

[13] Gunzburger, M.D.: Introduction to the mathematical aspects of flow control and optimization, Inverse design and optimization, von Karman Institute LS 1997-05, (1997).

[14] Ta'asan S.: Introduction to shape design and control, Inverse design and optimization, von Karman Institute LS 1997-5, (1997).

[15] Van den Braembussche, R.A., Léonard, O., Nekmouche, L.: Subsonic and transonic blade design by means of analysis codes, Computational Methods for Aerodynamic Design (Inverse) and Optimization, AGARD CP 463, Paper 9, (1989).

[16] Murugesan, K. and Railly, J.W.: Pure design method for aerofoils in cascade, J. Mechanical Engineering Science, Vol. 11, No 5, pp. 454-464, (1969).

[17] Sata, A., Ubaldi, M. and Zunino, P.: Design of axial turbines for mini hydro plants, Fourth International Symposium on Hydro Power Fluid Machinery, ASME FED-Vol. 43, pp. 29-36, (1984).

[18] Van den Braembussche, R.A.: The application of the singularity method to blade-to-blade calculations, Thermodynamics and Fluid Mechanics of Turbomachinery, NATO Advanced Sciences Institute Series, series E : Applied Sciences, No 97A, eds. A.S. Ucer, P. Stow and Ch. Hirsch, pp. 167-191, (1984).

[19] Arts, T.: Workshop on two- and three dimensional calculations in turbine bladings, Numerical Methods for Flows in Turbomachinery Bladings, von Karman Institute LS 1982-05, (1982).

[20] Sanger, N.L. and Shreeve R.P.: Comparison of calculated and experimental cascade performance for controlled diffusion compressor stator blading, ASME Journal of Turbomachinery, Vol. 108, pp. 43-50, (1986).

[21] Bogers P., Breugelmans F.A. and Van den Braembussche R.A.: Design and experimental verification of an optimized compressor blade, paper submitted for presentation at the ASME Gas Turbine conference, (1998).

[22] Léonard, O.: Subsonic and transonic cascade design, Inverse methods for Airfoil Design for Aeronautical and Turbomachinery Applications, AGARD-R-780, paper 7, (1990).

[23] Léonard, O.: Conception et développement d'une méthode inverse de type Euler et application à la génération de grilles d'aubes transsoniques, Ph.D. Thesis, Faculté Polytechnique de Mons & von Karman Institute, (1992).

[24] Léonard, O. and Van den Braembussche, R.A.: Design method for subsonic and transonic cascade with prescribed Mach number distribution, ASME Journal of Turbomachinery, Vol. 114, No 3, pp. 553-560, (1993).

[25] Léonard, O. and Van den Braembussche, R.A.: Inverse design of compressor and turbine blades at transonic flow conditions, ASME Paper 92-GT-430, (1993).

[26] Demeulenaere, A. and Van den Braembussche, R.A.: Inverse design of transonic blades taking into account radius change, von Karman Institute PR 1993-31, (1993).

[27] Van den Braembussche, R.A., Demeulenaere, A. and Borges, J.: Inverse design of radial flow impellers with prescribed velocity distribution, Technology Requirements for Small Gas Turbines, AGARD-CP-537, paper 18, (1993).

[28] Van den Braembussche R.A.: Inverse design methods for axial and radial turbomachines, Numerical methods for flow calculation in turbomachinery, von Karman Institute LS-1994-06, (1994).

[29] Demeulenaere, A. and Van den Braembussche, R.A.: Three-dimensional inverse method for turbomachinery blading design, ASME paper 96-GT-39, (1996).

[30] Demeulenaere, A. and Van den Braembussche, R.A.: Three-dimensional inverse design method for turbine and compressor blades, 3rd. International Symposium on Aerothermodynamics of Internal Flows, Beijing, September (1996).

[31] Meauzé, G.: An inverse time marching method for the definition of cascade geometry, ASME Journal of Engineering for Power, Vol. 104, pp. 650-656, (1982).

[32] Zannetti L. and Larocca F.: Inverse Method for 3D Internal Flows, Inverse Methods for Airfoil Design for Aeronautical and Turbomachinery Applications, AGARD-R- 780, paper 8, (1990).

[33] Demeulenaere, A. and Van den Braembussche, R.A.: A new compressor and turbine blade design method based on three-dimensional Euler computations with moving boundaries, VKI preprint

1997-56, (1997).

[34] Demeulenaere, A.: PhD thesis, University of Liege & von Karman Institute, (1997).

[35] Demeulenaere, A.: An Euler and Navier Stokes inverse method for compressor and turbine blade design, Inverse design and optimization, von Karman Institute LS 1997-5, (1997).

[36] Demeulenaere, A. and Van den Braembussche, R.: Application of a three-dimensional inverse method to the design of a centrifugal compressor impeller, 4th National Congress on Theoretical and Applied Mechanics, Leuven, pp. 353-356, (1997).

[37] Demirdzic, I. and Peric, M.: Space conservation law in finite volume calculations of fluid flow, International Journal for Numerical methods in Fluids, Vol. 8, pp. 1037-1050, (1988).

[38] Demirdzic, I. and Peric, M.: Finite volume method for prediction of fluid flow in arbitrarily shaped domains with moving boundaries, International Journal for Numerical methods in Fluids, Vol. 10, pp. 771-790, (1990).

[39] van Leer, B.: Flux vector splitting for the Euler equations, ICASE, Report No 82-30, (1982).

[40] Giles, M.: Non reflecting boundary conditions for Euler equations calculations, AIAA Journal, Vol. 108, No 12, pp 2050-2058, (1989).

[41] Arts, T., Lambert de Rouvroy, M. and Sieverding, C.H.: Highly loaded transonic linear turbine guide vane cascade LS89, in: Numerical Methods for flows in turbomachinery, von Karman Institute LS 1989-06, May 22-26, (1989).

[42] Han, W., Tan, C., Shi, H., Zhou, M. and Wang, Z.: Effects of leaning and curving of blades with high turning angles on the aerodynamic characteristics of turbine rectangular cascades, ASME Paper 93-GT-296, ( 1993).

[43] Léonard, O., Rogiest, P. and Delaneye M.: Blade analysis and design using an implicit flow solver, 2nd. European Conference on Turbomachinery-Fluid Dynamics and Thermodynamics, Antwerpen, March 5-7, pp. 331-338, (1997).

[44] Léonard, O. and Demeulenaere, A.: A Navier Stokes inverse method based on moving wall strategy, ASME paper 97-GT-416, (1997).

[45] Demeulenaere, A., Léonard, O. and Van den Braembussche R.: A two-dimensional Navier-Stokes inverse solver for compressor and turbine design, 2nd. European Conference on Turbomachinery-Fluid Dynamics and Thermodynamics, Antwerpen, March 5-7, pp. 339-346, (1997).

[46] Denton, J., Hodson, H.P. and Dominy R.G.: Subsonic turbine cascade LA, in: AGARD-AR-275, (1990).

[47] Deplaen, D. and Van den Braembussche R.A.: Ontwerp van transsone vleugelprofielen bij middel van een inverse Euler methode, von Karman Institute & Luru, afstudeerwerk, (1993).

# An Inverse Design Method for Wings Using Integral Equations and Its Recent Progress

Kisa Matsushima* and Susumu Takanashi** (f),


*   FUJITSU Ltd., 1-9-3, Nakase, Mihama-Ku, Chiba, 261, JAPAN
    E-mail kisam@nal.go.jp

    **   National Aerospace Laboratory
7-44-1, Jindaiji-Higashi-MachiChofu City, Tokyo, 182-8522, JAPAN

## Summary

An inverse design method using integral equations is explicated. Two other recently developed design methods are introduced. They are the extension of the first method. The formulation of an inverse problem used in each design method is also discussed for three major design categories. Each method designs a wing section shape which realizes the prescribed target pressure distribution by iterating a residual-correction loop which consists of a flow simulation and an inverse problem. It starts with the initial guess of current wing shape. The residual, defined as the difference between the target and simulated current pressure distributions, is compensated by solving the inverse problem. The inverse problem determines the section geometry of the wing. Each of three inverse problems here are formulated to be an integral equation system by mathematically converting the partial differential equations which govern the flowfield. The first inverse problem is for transonic wing design. The second one is for supersonic and the third one is for design of multiple wing systems. Emphasis is put on the discussion of the formulation. Works on wing design using the method with the first inverse problem are cited. Design problems by the second and third ones are also presented.

## Introduction

Automatic aerodynamic design of aircraft has been one of the major applications of Computational Fulid Dynamics (CFD). In aerodynamics, there are two traditional computational design techniques. One is the inverse problem technique. The other is numerical optimization which is formed by numerous flow simulations with any of numerical algorithm of searching for the optimum shape. Each method has advantages and drawbacks when compared with the other. From the viewpoint of engineering application, a technique providing a faster design cycle gives a crucial advantage. In this sense, the inverse problem technique is desirable because it requires a much smaller number of flow simulations than the numerical optimization does. The inverse problem technique recognizes that the designer usually has an

179

idea of the kind of pressure distribution that will lead to the desired performance.

A considerable number of inverse design methods have been devised, so far[1,2]. Among them, the method using integral equations is regarded as one of the most versatile and efficient design methods. The method was originally developed for a wing in transonic flows in 1985 by one of the authors, the late Dr. Takanashi[3]. He adopted residual-correction design methodology for the method, which is outlined in the next section. With this method, he formulated the mathematical inverse problem where the aerodynamic geometry was obtained as the solution of integral equations. They were derived by converting the differential equations of physics in the flowfield as shown in the third and forth sections. Thus, the integral equations are the mathematical model to describe the physical relation of the wing section geometry to the surface pressure distribution. Much interesting work has been done on the original integral equation method[4-10]. Works will be cited, later.

In this article, the background and the theory of the formulation are explained in greater detail than the usual articles, for theoretical bases and assumptions should be useful to readers, especially those who would intend to learn the inverse problem. Therefore, a sizable portion of this article deals with the formulation process rather than the results.

Recently, two projects to expand the original method have started. One is for supersonic flows; the other is for multiple wing systems with strong interaction. For each of them, a new inverse problem has been formulated. Both of them have accomplished useful design results. Four sections are spent with the first project. Three sections are devoted to the second project.

Before concluding the introduction, the authors would like to strictly define the usage of the term, "an inverse problem". Without the definition, the discussion of this article would be confusing. In aerodynamics, "an inverse problem" is used in two ways, aerodynamically and mathematically. Aerodynamically, it means to determine an airfoil/wing shape which produces the prescribed target pressure distribution. Mathematically, it means to solve mathematical equations which are derived from the differential equations representing the physics of the situation. In this article, we will refer to the first interpretation as "inverse design" or "an inverse method", while "a mathematical inverse problem" or "an inverse problem" is used to mean the other interpretation.

## Outline of the Design Method – Iterative Procedure –

The design procedure is iterative. Fig.1 illustrates the procedure. The goal is to determine the wing section geometry which realizes a specified target pressure distribution at all span stations of a wing. The goal can be attained asymptotically by iterating a "residual-correction" loop. The residual is defined as the difference between the target and current pressure distributions. The correction is made on the wing geometry to make up for the residual. To be more specific, the procedure starts with a baseline shape. First, the flowfield around the baseline wing is analyzed by flow simulation to get the current $Cp$ distribution on the wing surface. Next the inverse problem is solved to obtain the geometrical correction value, $\Delta f$, corresponding to the difference between target and current pressure distributions, $\Delta Cp$. The new wing is designed by modifying the baseline shape using $\Delta f$. Now, the current shape is updated. The next step is to go back to the flowfield analysis. The flow analysis is conducted to see if the current shape realizes target pressure distribution. If the difference between target and current pressure distributions is negligible, the design is complete. Otherwise, the next step is once again to solve the inverse problem and iterate the residual-correction loop until the pressure difference becomes negligible.

This procedure has two primary parts; one is flow analysis, where grid generation and flow simulation are conducted. Flow simulation is sometimes referred to as a direct problem, because it solves differential equations which directly describe physical phenomena. The other is an inverse design part where the mathematical inverse problem, i.e. the integral equation system, is solved to update the wing geometry. Both parts are completely independent from each other. The analysis part evaluates the residual. The design part provides the correction value for asymptotic approach to a solution. Thus, the method may be called either a residual-correction method or a direct-inverse coupling method. For the design part, efficient methods using integral equations are going to be introduced in the following sections. For the analysis part, any kind of simulation code can be employed or even a wind tunnel test can replace the computational analysis, as long as it provides a reasonably accurate pressure distribution on the wing surface. In the first paper presented on this method in 1985[3], a potential flow simulation code was used, such as FLO-22. In later references [4-10], a variety of flow simulation codes have been employed with this method.

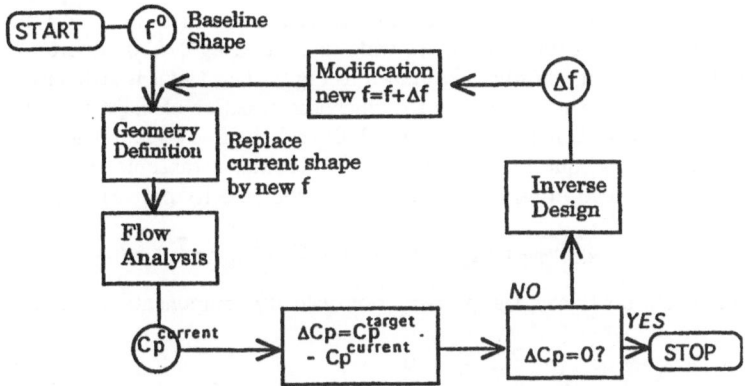

Fig. 1 Iterative procedure of design method.

Since the residual is evaluated in the analysis part, the accuracy of the designed geometry by this method depends on the flow simulation code. The basic equations for the design part come from the small disturbance equation and the thin wing theory. But this doesn't restrict the design results, as readers will see in the next section. When a Navier-Stokes flow simulation is conducted and the iterations converge, the design result is accurate in Navier-Stokes flows regardless of what the flow equations of the inverse design part are. In this case, the small disturbance equation is assumed to be the approximation of the perturbation of the Navier-Stokes equations. If the variation of the viscous effect and the rotation of the flowfield caused by the perturbation is small enough, that approximation is quite appropriate. However, the assumption sometimes might lose validity, as for example, in the case where strong separation would be caused by the perturbation. We have to be careful about the assumption for the approximation. If the assumption is inappropriate for the design problem, we will never reach the solution. If the design part solved the inverse problem of the Navier-Stokes equations, it would be more desirable. It is true. Unfortunately, so far, no practical formulation of the mathematical inverse problem for the Euler or Navier-Stokes equations has been successfully performed. In addition, from the viewpoint of engineering application, too much computational cost and complexity should be avoided. Hence, the inverse problem of the small disturbance equation will be considered in the following sections. The small disturbance equation is much simpler than the Euler and Navier-Stokes equations and Green's theorem can easily be applied to it to formulate the mathematical inverse problem.

## Formulation I
### $\Delta$-form Equations for Residual-Correction Concept

The basis of this aerodynamic shape design is to construct mathematical equations to relate geometrical correction to the difference between two pressure distributions. In this section, aerodynamic equations to describe the difference between two states of a flowfield are derived. The formulation starts with the well-known small disturbance approximation. A wing is located at $\bar{z} = 0$ with an angle of attack of 0 degrees in a flowfield as shown in Fig 2. The $\bar{x}$ axis is streamwise, the $\bar{y}$ axis is spanwise and the $\bar{z}$ axis is in the thickness direction of the wing. The undisturbed flow velocity is normalized as $(1, 0, 0)$ and the free-stream Mach number is $M_\infty$. The governing equation applicable on a subsonic, a transonic and a supersonic flowfields is described in terms of the perturbation velocity potential, $\phi$, as[11]

$$(1 - M_\infty^2)\bar{\phi}_{\bar{x}\bar{x}} + \bar{\phi}_{\bar{y}\bar{y}} + \bar{\phi}_{\bar{z}\bar{z}} = (\gamma + 1)M_\infty^2 \frac{\partial}{\partial \bar{x}}(\frac{1}{2}\bar{\phi}_{\bar{x}}^2) + h.o.t. \tag{1}$$

At the far-field boundary, the perturbation velocity components are all zero; that is
When $(\bar{x}, \bar{y}, \bar{z}) \to \infty$, $(\bar{\phi}_{\bar{x}}, \bar{\phi}_{\bar{y}}, \bar{\phi}_{\bar{z}}) \to 0$.
At the wing surface, the flow is tangential to the surface shape. This tangency condition is described as

$$\bar{\phi}_{\bar{z}}(\bar{x}, \bar{y}, \pm 0) = \frac{\partial}{\partial \bar{x}}\bar{f}_{\pm}(\bar{x}, \bar{y}) + h.o.t. \tag{2}$$

The geometry of the upper and lower surface of the wing is expressed as $\bar{z}_{\pm} = \bar{f}_{\pm}(\bar{x}, \bar{y})$, where $+$ and $-$ indicate the upper and the lower surfaces, respectively.
The pressure coefficients on the wing surface are related to the perturbation velocity as follows;

$$Cp_{\pm}(\bar{x}, \bar{y}) = -2\bar{\phi}_{\bar{x}}(\bar{x}, \bar{y}, \pm 0) + h.o.t. \tag{3}$$

Eq.(1) is the nonlinear small disturbance equation with higher order terms of the perturbation quantity. Eqs.(2) and (3) are the linearized boundary conditions with higher order terms. They can also be regarded as errors from the more accurate (less approximated) equations.

Then, another state of the flowfield is considered. We introduce the increment in the perturbation velocity potential, $\Delta\phi$, due to the geometrical change from the original state, $\Delta f$. The governing equation becomes

$$(1 - M_\infty^2)(\overline{\phi}_{\overline{x}\overline{x}} + \Delta\overline{\phi}_{\overline{x}\overline{x}}) + (\overline{\phi}_{\overline{y}\overline{y}} + \Delta\overline{\phi}_{\overline{y}\overline{y}}) + (\overline{\phi}_{\overline{z}\overline{z}} + \overline{\phi}_{\overline{z}\overline{z}}) =$$

$$\frac{\partial}{\partial\overline{x}}\left[(\frac{1}{2}\overline{\phi}_{\overline{x}} + \Delta\overline{\phi}_{\overline{x}})^2\right] + h.o.t + \Delta(h.o.t) . \qquad (4)$$

The tangency condition with the increment can be written as

$$\overline{\phi}_{\overline{z}}(\overline{x},\overline{y},\pm0) + \Delta\overline{\phi}_{\overline{z}}(\overline{x},\overline{y},\pm0) = \frac{\partial}{\partial\overline{x}}[f(\overline{x},\overline{y}) + \Delta f(\overline{x},\overline{y})] + h.o.t + \Delta(h.o.t) . \qquad (5)$$

The pressure on the wing surface varies to

$$Cp_\pm(\overline{x},\overline{y}) + \Delta Cp_\pm(\overline{x},\overline{y}) = -2[\overline{\phi}_{\overline{x}}(\overline{x},\overline{y},\pm0) + \Delta\overline{\phi}_{\overline{x}}(\overline{x},y,\pm0)] + h.o.t + \Delta(h.o.t). \qquad (6)$$

If the potential increment $\Delta\overline{\phi}(\overline{x},\overline{y},\overline{z})$ is small and the flowfield is not too sensitive to the disturbance, the increment of higher order terms, $\Delta(h.o.t)$, may be assumed negligibly small even when shock waves are present. Therefore, we get the equations for the increment, $\Delta\overline{\phi}$, $\Delta\overline{f}_\pm$ and $\Delta Cp_\pm$.

$$(1 - M_\infty^2)\Delta\overline{\phi}_{\overline{x}\overline{x}} + \Delta\overline{\phi}_{\overline{y}\overline{y}} + \Delta\overline{\phi}_{\overline{z}\overline{z}} = \frac{\partial}{\partial\overline{x}}\left[\frac{1}{2}(\overline{\phi}_{\overline{x}} + \Delta\overline{\phi}_{\overline{x}})^2 - \frac{1}{2}\overline{\phi}_{\overline{x}}^2\right] \qquad (7)$$

$$\Delta\overline{\phi}_{\overline{z}}(\overline{x},\overline{y},\pm0) = \frac{\partial}{\partial\overline{x}}\Delta\overline{f}_\pm(\overline{x},\overline{y}) \qquad (8)$$

$$\Delta Cp_\pm(\overline{x},\overline{y}) = -2\Delta\overline{\phi}_{\overline{x}}(\overline{x},\overline{y},\pm0) . \qquad (9)$$

Now, the basic equations for the mathematical inverse problem are converted into a $\Delta$-form, which corresponds to the residual-correction concept of the design method. These above $\Delta$-form equations may be regarded as the approximated equations for the difference between two Navier-Stokes, Euler or potential flowfields. It depends what one puts in $h.o.t$. The formulation starts from the small disturbance equation, but this does not restrict the applicability of the method. This inverse problem can be used with a Navier-Stokes, Euler, or potential flow analysis. The $\Delta$-form formulation is useful for many kinds of aerodynamic design problems as long as one is aware of the omitted terms, $\Delta(h.o.t)$.

Here, we need to introduce the assumption for the flowfield that $M_\infty < 1.0$. The rest of this section and the next section are going to be concerned with a transonic flowfield.

Applying the Prandtl-Glauert transformation such as

$x = \overline{x}, \; y = \beta\overline{y}, \; z = \beta\overline{z}, \; \phi(x,y,z) = \frac{K}{\beta^2}\phi(\overline{x},\overline{y},\overline{z}), \; \text{and} \; f_\pm(x,y) = \frac{K}{\beta^3}\overline{f}_\pm(\overline{x},\overline{y})$

where $\beta = \sqrt{1 - M_\infty^2}$ and $K = (\gamma+1)M_\infty^2$.

Eqs. (7-9) yield

$$\Delta\phi_{xx} + \Delta\phi_{yy} + \Delta\phi_{zz} = \frac{\partial}{\partial x}\Big[\frac{1}{2}(\phi_x + \Delta\phi_x)^2 - \frac{1}{2}\phi_x^2\Big] \tag{10}$$

$$\Delta\phi_z(x, y, \pm 0) = \frac{\partial}{\partial x}\Delta f_\pm(x, y) \tag{11}$$

$$\Delta Cp_\pm\Big(x, \frac{y}{\beta}\Big) = -2\frac{\beta^2}{K}\Delta\phi_x(x, y, \pm 0) \ . \tag{12}$$

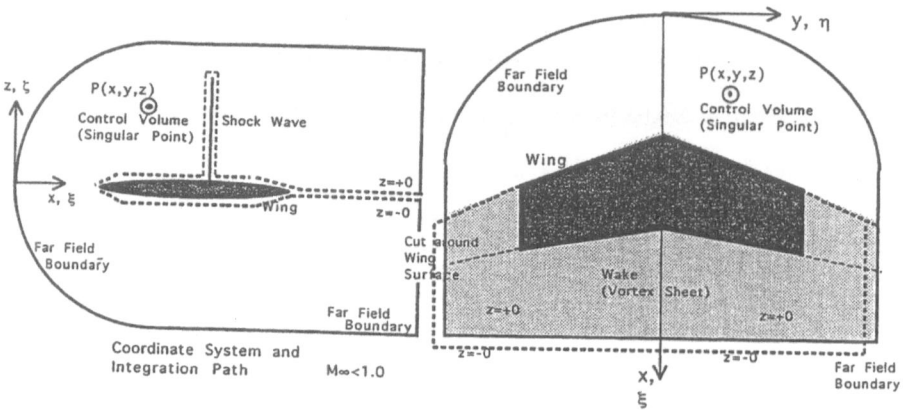

Fig. 2 Flowfield and coordinate system for formulation ($M_\infty < 1.0$).

## Formulation II
### An Inverse Problem for Transonic Flows

In this section, the differential equations are transformed to integral equations. First, Green's theorem is applied to Eq.(10). The corresponding Green's theorem to Eq.(10) is

$$\int\int\int_V [\Psi L(\Delta\phi) - \Delta\phi L(\Psi)]dV = -\int\int_\Sigma \Big(\Psi\frac{\partial\Delta\phi}{\partial\bar{n}} - \phi\frac{\partial\Psi}{\partial\bar{n}}\Big)d\Sigma \tag{13}$$

where $L(\Delta\phi)$ is defined as

$$L(\Delta\phi) \equiv \Delta\phi_{xx} + \Delta\phi_{yy} + \Delta\phi_{zz}$$

and $\frac{\partial\Delta\phi}{\partial\bar{n}}$ is a derivative along the conormal of the surface $\Sigma$. The fundamental solution of the differential equation, $L(\Psi) = 0$, is now selected as the arbitrary function $\Psi$ in Eq.(13). The following relations hold:

$$\int\int_\Sigma \Big(\Psi\frac{\partial\Delta\phi}{\partial\bar{n}} - \phi\frac{\partial\Psi}{\partial\bar{n}}\Big)d\Sigma = -\int\int\int_V \Psi L(\Delta\phi)dV$$

$$= -\int\int\int_V \Psi\frac{\partial}{\partial x}\Big[\frac{1}{2}(\phi_x + \Delta\phi_x)^2 - \frac{1}{2}\phi_x^2\Big] \tag{14}$$

where,

$$\Psi = \frac{1}{\sqrt{(x-\xi)^2 + (y-\eta)^2 + (z-\zeta)^2}} \ . \tag{15}$$

The domain of Green's integration, $V$, is indicated in Fig.2. The running coordinates for the integration are $\xi$, $\eta$ and $\zeta$. The control volume, where $\Delta\phi$ is to be evaluated, is also shown in Fig.2. The control volume is a sphere of infinitesimal radius and its center is $P(x, y, z)$ where $\Psi$ becomes singular. To maintain the mathematical correctness of the formulation, the domain is carefully defined so that $\Delta\phi$ should be $C^1$ continuous. The domain is surrounded by the boundary surfaces, $\Sigma$. $\Sigma$ consists of an infinite far-field boundary which is the external boundary of $V$, the sphere surface of a control volume and a surface enveloping the wing, its wake and its shock waves.

At an infinite far-field boundary, the disturbance associated with small changes in wing geometry vanishes, that is, $\Delta\phi$ and the velocity change (the partial derivative of $\Delta\phi$ ) sould be zero. The surface integrals over a far-field boundary have no contribution to Eq.(14). The surface integrals over the sphere of a control volume result in $4\pi\Delta\phi(x, y, z)$. The contribution of the integrals along the envelope of shock waves vanishes due to the shock-polar relationships[11], assuming that they are all normal to the wing surface. In addition, performing integration by parts and applying the divergence theorem to the volume integral term, Eq.(14) yields

$$\Delta\phi(x,y,z) = -\frac{1}{4\pi} \int\int_{S_{wv}} \Psi(x,y,z;\xi,\eta,0)\big[\Delta\phi_\zeta(\xi,\eta,+0) - \Delta\phi_\zeta(\xi,\eta,-0)\big]d\xi d\eta$$

$$+\frac{1}{4\pi} \int\int_{S_{wv}} \Psi_\zeta(x,y,z;\xi,\eta,0)\big[\Delta\phi(\xi,\eta,+0) - \Delta\phi(\xi,\eta,-0)\big]d\xi d\eta$$

$$+ \frac{1}{4\pi} \int_{-\infty}^{\infty}\int_{-\infty}^{\infty}\int_{-\infty}^{\infty} \Psi_\xi(x,y,z;\xi,\eta,\zeta)\chi(\xi,\eta,\zeta)d\xi d\eta d\zeta \tag{16}$$

where

$$\chi(x,y,z) = 1/2[(\phi_x + \Delta\phi_x)^2 - \phi_x^2] \tag{17}$$

$$\Psi(x,y,z;\xi,\eta,\zeta) = \frac{1}{\sqrt{(x-\xi)^2 + (y-\eta)^2 + (z-\zeta)^2}} \ . \tag{18}$$

In the first and second terms, the domain of the surface integral denoted by $S_{wv}$ means the mean plane of a wing and its wake. We assume the thin wing theory so that the $S_{wv}$ plane is parallel with the plane of $z =$constant. The direction cosines of the normal to the surface of $S_{wv}$, $\bar{n}$, are $(0, 0, 1)$. Then, $\frac{\partial}{\partial\bar{n}}$ becomes $\frac{\partial}{\partial\zeta}$. $+0$ and $-0$ indicate the upper and lower surfaces of the $S_{wv}$ plane, respectively.

To evaluate the volume integral in Eq.(16 ), the finite part integral such as

$$\int_{-\infty}^{\infty}\int_{-\infty}^{\infty}\int_{-\infty}^{\infty}(\ )d\xi d\eta d\zeta = \lim_{\epsilon \to 0}\int_{-\infty}^{\infty}\int_{-\infty}^{\infty}[\int_{-\infty}^{x-\epsilon}(\ )d\xi + \int_{x+\epsilon}^{\infty}(\ )d\xi]d\eta d\zeta \tag{19}$$

is adopted[3,11], since the integrand is singular at the point $(\xi, \eta, \zeta) = (x, y, z)$.

To combine Eq.(14) with Eqs.(11) and (12) which describe geometry and pressure distribution respectively, further calculus is needed. Differentiating both sides of Eq.(16) with respect to $x$ and adding the resulting equation of $\Delta\phi_x(x,y,z)$ at $z = +0$ to that at $z = -0$, we obtain

$$\Delta u_s(x,y) = -\frac{1}{2\pi} \int\int_{S_w} \Psi(x,y,0;\xi,\eta,0)\Delta w_s(\xi,\eta)d\xi d\eta + \chi_s(x,y)$$

$$+ \frac{1}{4\pi} \int\int\int_{-\infty}^{\infty} \Psi_{\xi x}(x,y,0;\xi,\eta,\zeta)[\chi(\xi,\eta,\zeta) + \chi(\xi,\eta,-\zeta)]d\xi d\eta d\zeta \qquad (20)$$

where

$$\chi_s(x,y) = \chi(x,y,+0) + \chi(x,y,-0) \qquad (21)$$

$$\Delta u_s(x,y) = \Delta\phi_x(x,y,+0) + \Delta\phi_x(x,y,-0) \qquad (22)$$

$$\Delta w_s(x,y) = \Delta\phi_z(x,y,+0) - \Delta\phi_z(x,y,-0) . \qquad (23)$$

The second term, $\chi_s(x,y)$, is generated from the improper integral after differentiating the finite part integration of the volume integral in Eq.(16)[11].

Next, differentiating both sides of Eq.(16) with respect to $z$, adding the value of $\Delta\phi_z(x,y,z)$ at $z = +0$ to that at $z = -0$, and using an integral by parts, we obtain

$$\Delta w_a(x,y) = \frac{1}{2\pi} \int\int_{S_w} \frac{\Delta u_a(\xi,\eta)}{(y-\eta)^2}\Big[1 + \frac{x-\xi}{\sqrt{(x-\xi)^2 + (y-\eta)^2}}\Big]d\xi d\eta$$

$$+ \frac{1}{4\pi} \int\int\int_{-\infty}^{\infty} \Psi_{\xi z}(x,y,0;\xi,\eta,\zeta)[\chi(\xi,\eta,\zeta) - \chi(\xi,\eta,-\zeta)]d\xi d\eta d\zeta \qquad (24)$$

where

$$\Delta u_a(x,y) = \Delta\phi_x(x,y,+0) - \Delta\phi_x(x,y,-0) \qquad (25)$$

$$\Delta w_a(x,y) = \Delta\phi_z(x,y,+0) + \Delta\phi_z(x,y,-0) . \qquad (26)$$

In order to evaluate the volume integrals of Eqs. (20) and (24), a velocity decay function

$$\phi_x(x,y,z) = \phi_x(x,y,\pm 0)\exp[\mp R_\pm(x,y)z] \qquad (27)$$

where

$$R_\pm(x,y) = \Big|\frac{\partial^2}{\partial x^2}f_\pm(x,y)/\phi_x(x,y,\pm 0)]\Big| \qquad (28)$$

is introduced to approximate the exponential decay of the perturbation velocity component $\phi_x$ in the $z$ direction. This decay function was introduced by Nørstud[12] for analyzing transonic flows past a wing. It is used here for approximating the small change in the perturbation velocity component along the $z$ direction

$$\Delta\phi_x(x,y,z) = \Delta\phi_x(x,y,\pm 0)\exp[\mp R_\pm(x,y)z] . \qquad (29)$$

Thus, $\chi(x,y,z)$ is expressed as

$$\chi(x,y,z) = \frac{1}{2}\Big\{[\phi_x(x,y,z) + \Delta\phi_x(x,y,z)]^2 - \phi_x^2(x,y,z)\Big\}$$

$$= \frac{1}{2}\Big\{[\phi_x(x,y,\pm 0) + \Delta\phi_x(x,y,\pm 0)]^2 - \phi_x^2(x,y,\pm 0)\Big\}\exp[\mp 2R_\pm(x,y)z]$$

$$= \chi(x, y, \pm 0) \exp[\mp 2R_\pm(x, y)z] \ . \tag{30}$$

Furthermore, assuming that $\chi(x, y, \pm 0) = 0$ unless a point $P(x, y, \pm 0)$ is on the wing surface and substituting Eq. (30) to Eqs. (20) and (24), Eq. (20) yields

$$\Delta u_s(x, y) = -\frac{1}{2\pi} \int \int_{S_w} \Psi_x(x, y, 0; \xi, \eta, 0) \Delta w_s(\xi, \eta) d\xi d\eta + \chi_s(x, y)$$

$$+ \frac{1}{2\pi} \int \int_{S_w} [I_{s+}(x, y, ; \xi, \eta)\chi(\xi, \eta, +0) + I_{s-}(x, y, ; \xi, \eta)\chi(\xi, \eta, -0)] d\xi d\eta. \tag{31}$$

Eq. (24) yields

$$\Delta w_a(x, y) = \frac{1}{2\pi} \int \int_{S_w} \frac{u_a(\xi, \eta)}{(y - \eta)^2} \left[1 + \frac{x - \xi}{\sqrt{(x - \xi)^2 + (y - \eta)^2}}\right] d\xi d\eta$$

$$+ \frac{1}{2\pi} \int \int_{S_w} \left[I_{a+}(x, y, ; \xi, \eta)\chi(\xi, \eta, +0) - I_{a-}(x, y, ; \xi, \eta)\chi(\xi, \eta, -0)\right] d\xi d\eta \tag{32}$$

where

$$I_{s\pm}(x, y; \xi, \eta) = \int_0^\infty \Psi_{\xi x}(x, y, 0; \xi, \eta, \zeta) \exp[-2R_\pm(\xi, \eta)\zeta] d\zeta \tag{33}$$

$$I_{a\pm}(x, y; \xi, \eta, \pm 0) = \int_0^\infty \Psi_{\xi z}(x, y, 0; \xi, \eta, \zeta) \exp[-2R_\pm(\xi, \eta)\zeta] d\zeta \ . \tag{34}$$

As the reader will notice, Eqs. (33) and (34) are the equations of the Laplace transform; $I_{s\pm}$ and $I_{a\pm}$ can be analytically calculated using the Neumann function as well as the Struve function[12]. Alternatively, numerical integration can be applied, which is much simpler than the analytical evaluation. In the original paper of the design method[3], a numerical integration technique was used.

The three-dimensional inverse problem is consequently reduced to rather simple two-dimensional integral equations. The integration now remains only on the mean plane surface of the wing, $S_w$. The unknowns of the above equations are $\Delta w_s$ for Eq. (31) and $\Delta w_a$ for Eq. (32). These functions have physical meaning related to the geometrical correction of the wing surface shape as follows. Substituting Eq. (11) into Eqs. (23) and (26), the unknown functions are expressed as

$$\Delta w_s(x, y) = \frac{\partial}{\partial x} \Delta f_s(x, y) \tag{35}$$

and

$$\Delta w_a(x, y) = \frac{\partial}{\partial x} \Delta f_a(x, y) \tag{36}$$

where

$$\Delta f_s(x, y) = \Delta f_+(x, y) - \Delta f_-(x, y) \tag{37}$$

$$\Delta f_a(x, y) = \Delta f_+(x, y) + \Delta f_-(x, y). \tag{38}$$

$\Delta f_s(x, y)$ represents the geometrical correction value symmetric with respect to the mean plane of the wing, in other words, correction in thickness of the wing. $\Delta f_a(x, y)$ represents the anti-symmetric correction value, in other words, correction in curvature of the camber of the wing.

One of the unknown functions, $\Delta w_a$, is determined by straightforward evaluation of the right-hand side of Eq. (32). However, the other unknown function $\Delta w_s$, which appears implicitly in Eq.(31), is determined by solving the integral equation. According to the theory concerning an integral eqation having the Cauchy kernel, an additional condition has to be imposed on the unknown function to let an integral equation such as Eq.(31) be definitely solvable. The following condition is imposed on $\Delta w_s$ at any $y$ coordinate:

$$\int_{L.E.}^{T.E.} \Delta w_s(x, y) dx = 0 \,. \tag{39}$$

The above condition is important not only mathematically but also physically. Mathematically, it assures the uniqueness of the solution. Physically, it assures that every section of a designed wing by this inverse problem has a closed trailing-edge provided the initial wing had one. Eq.(39) implies that the trailing-edge thickness of a designed wing is as same as that of the initial wing. Eventually, the geometrical correction is calculated using

$$\Delta f_{\pm}(x, y) = \frac{1}{2} \int_{L.E.}^{x} \left[ \Delta w_s(\xi, y) \pm \Delta w_a(\xi, y) \right] d\xi \,. \tag{40}$$

## Piece-wise Function Approximation

In order to facilitate the evaluation of integrals of Eqs.(31) and (32), the domain of integrals is discretized. As shown in Fig.3, the wing surface, $S_w$ is divided into small rectangular panels. In the spanwise ($y$) direction, there are $2J + 1$ intervals. In the chordwise direction, there are $I$ intervals. Thus, the total number of panels is $I(2J + 1)$.

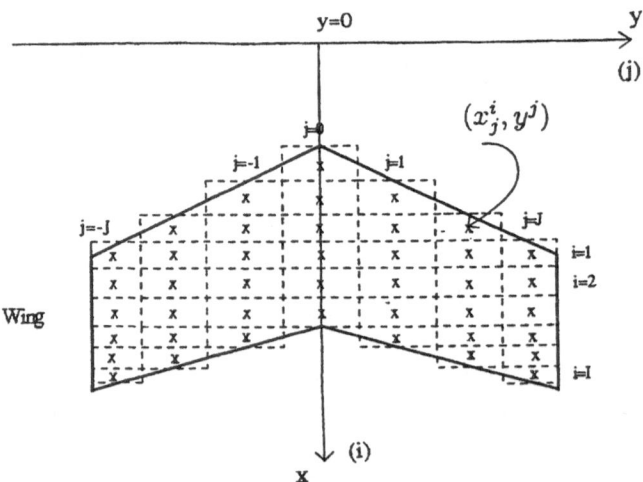

Fig. 3 Panels on a wing surface.

On each panel, $\Delta u_s(x, y)$, $\Delta u_a(x, y)$, $\chi(x, y, \pm 0)$, $R_{\pm}(x, y)$, and $\Delta w_a(x, y)$ are assumed to be constant. The values of those functions are defined at the center of a panel, which is denoted by $(x_j^i, y^j)$. On the other hand, $\Delta w_s(x, y)$ is assumed to vary linearly in the $x$ direction but to be constant in the $y$ direction on each panel. The value of $\Delta w_s$ is defined at the middle point of a $y$-direction side of a panel, which is denoted by $(x_j^{i-1/2}, y^j)$. With these piecewise function approximations, the surface integrals of Eqs. (31) and (32) are transformed to the summation of integrals inside every panel. The integrals inside a panel can be analytically evaluated.

Recalling the symmetry of a flowfields about $y = 0$, the discretized equations of Eqs. (31) and (32) are

$$\Delta u_s(x_j^i, y^j) = \sum_{k=1}^{I+1} \sum_{m=0}^{J} \mu_{ijkm}^s \Delta w_s(x_m^{k-1/2}, y^m) + \chi_s(x_j^i, y^j)$$

$$+ \sum_{k=1}^{I} \sum_{m=0}^{J} \left[ \nu_{ijkm}^s \chi(x_m^k, y^m, +0) + \tilde{\nu}_{ijkm}^s \chi(x_m^k, y^m, -0) \right] \qquad (41)$$

$$\Delta w_a(x_j^i, y^j) = \sum_{k=1}^{I+1} \sum_{m=0}^{J} \mu_{ijkm}^a \Delta u_a(x_m^k, y^m)$$

$$+ \sum_{k=1}^{I} \sum_{m=0}^{J} \left[ \nu_{ijkm}^a \chi(x_m^k, y^m, +0) - \tilde{\nu}_{ijkm}^a \chi(x_m^k, y^m, -0) \right]. \qquad (42)$$

Where the coefficients $\mu_{ijkm}^s$, $\mu_{ijkm}^a$, $\nu_{ijkm}^s$, $\tilde{\nu}_{ijkm}^s$, $\nu_{ijkm}^a$ and $\tilde{\nu}_{ijkm}^a$ are the results of the corresponding piecewise integrations inside the panel of (k,m).

Eq.(41) is the linear equation system for $\Delta w_s$. At each span station, the number of unknowns to define $\Delta w_s(x, y)$ is $I+1$, since the number of $y$-direction sides there is $I+1$. However, Eq.(41) provides $I$ equations. To regularize the equation system, an additional equation at each spanstation such as Eq.(39) has to be introduced. The discretized form of Eq.(39) at any $j$ station is

$$\sum_{i=1}^{I} \frac{1}{2} \left[ \Delta w_s(x_j^{i-1/2}, y^j) + w_s(x_j^{i+1/2}, y^j) \right] (x_j^{i+1/2} - x_j^{i-1/2}) = 0 \quad (j = 0, 1, ...J). \qquad (43)$$

Thus, the linear equation system for $\Delta w_s$ is mathematically guaranteed to have a unique solution. The linear equation system can be solved by standard techniques such as Gaussian elimination.

## Remarks on Basic Equations for the Inverse problem

In this section, the background of two equations, Eqs.(2) and (3), which augment the small disturbance equation, Eq.(1), is briefly explained. To formulate the integral equations of the inverse problem, the small disturbance equation and Green's theorem take the primary role. Afterwards, the other two equations also take an important role to construct the relationship between the pressure difference and the geometrical correction. Eqs.(2) and (3) come from the thin wing theory. Since CFD is now enjoying the advantage of the Navier-Stokes equations, the thin wing theory is less commonly used in CFD. It is important to recall the theory and make clear what the two equations imply, especially when one is trying to rate the applicability of the design method and improve it.

The flowfield to be considered is the same as that for the formulation I. A wing is located in the flowfield where the free-stream velocity is $(1,0,0)$ and the free-stream Mach number is $M_\infty$. The wing is thin and located on the $x$-$y$ plane at $z = 0$ as shown in Fig.2. The thin wing theory regards the upper and lower surface locations of the wing as the perturbation from the mean plane of the wing. So, they are at $z = +\epsilon$ and $z = -\epsilon$, where $\epsilon \ll 1$. The perturbation in velocity is $(u, v, w)$. We assume the disturbance by the wing should be small, so the perturbation velocity components, $u, v$, and $w$, are $\ll 1$.

## Approximation of pressure coefficients

Pressure coefficients are defined as

$$C_p = \frac{p - p_\infty}{1/2\rho_\infty} = \frac{2}{\gamma M_\infty^2} \frac{p - p_\infty}{\rho_\infty} . \tag{44}$$

In terms of perturbation velocities, u, v, and w, it is also expressed as

$$C_p = \frac{2}{\gamma M_\infty^2}\left\{\left[1 - \frac{\gamma - 1}{2} M_\infty^2(2u + u^2 + v^2 + w^2)\right]^{\frac{\gamma}{(\gamma-1)}} - 1\right\}. \tag{45}$$

Applying binomial expansion to the term of $(1 - \frac{\gamma-1}{2}M_\infty^2...)^{\frac{\gamma}{(\gamma-1)}}$ with respect to $u$,

$$Cp = -2u - (1 - M_\infty^2)u^2 - v^2 + w^2 \tag{46}$$

is obtained in the second order approximation. When one is handling the pressure distribution on the wing surface which lies parallel to the free-stream flow direction as shown in Fig.2, the second order terms may be neglected with the assumption of small disturbance. Thus, the linearized pressure equation

$$Cp = -2u \tag{47}$$

is derived. Assuming that both the upper and lower wing surfaces are approximately on the mean plane of the wing, which is the plane of $z = 0$, the pressure coefficient at the both wing surfaces is expressed as

$$Cp_\pm(x, y) = -2 \lim_{\epsilon \to 0} \phi_x(x, y, \pm\epsilon) = -2\phi_x(x, y, \pm 0) \tag{48}$$

where $+$ and $-$ indicate the upper and lower sufaces, respectively.

## Approximation of tangency (slip) condition

When a flow can be regarded as inviscid, a flow in the vicinity of the wing surface streams tangentially to the surface shape; that is, the velocity vector on the surface must be perpendicular to the normal of the wing surface plane. The shape functions for the surface of the wing are denoted by $f(x, y)$. The equation of the wing surface plane is $z = f(x, y)$. The normal vector of the plane is

$$\vec{n} = (-\frac{\partial f(x, y)}{\partial x}, -\frac{\partial f(x, y)}{\partial y}, 1) .$$

The velocity vector of the perturbed flowfield is $\vec{v} = (1 + u, v, w)$. Then, the inner product of both vectors becomes zero;

$$- (1 + u)f_x - vf_y + w = 0 \tag{49}$$

$u, v$ and $w \ll 1$. The order of $f_x$ and $f_y$ is $\epsilon$; $\epsilon$ is the wing thickness normalized by the standard length of the flowfield, such as the whole length of an airplane or the wing chord length at a root section. We assume $\epsilon \ll 1$ in the thin wing theory. Omitting the second order terms of the perturbation quantity in Eq.(49), the relation

$$(-f_x + w)_{surface} = 0 \tag{50}$$

is derived. Since the approximate wing surface location is the plane of $z = \pm\epsilon \approx 0$, Eq.(50) can be written as

$$\phi_z(x, y, \pm 0) = \frac{\partial}{\partial x} f_\pm(x, y) \ . \tag{51}$$

## Works with the Design Method

Many interesting subjects in research and design have been explored with the design method, since the issue of the first paper of the original work in 1985[3]. They are listed in References[4−10]. The inverse problem shown in the article is of precious value. It was formulated three-dimensionally at a time when such inverse problems that could handle three-dimensional design was scarce. Such inverse problems are still scarce. Hirose et. al. confirmed the feasiblity of design in Navier-Stokes flows using the inverse problem[4]. Xia et. al. implemented their own software to solve the inverse problem and applied it to the design of a transonic wing[5]. Lorentzen formulated a two dimensional inverse problem based on the formulation of the original work presented in this article[6]. Hua et. al. developed an efficient transonic wing of a new concept with the design method[7]. They employed a potential flow solver fortified by boundary layer analysis code for the flow analysis part. The design result was successfully assured by wind-tunnel testing. Bartelheimer improved the robustness of the inverse problem and presented a good design result for a particular transonic design[8]. Obayashi combined the method with the genetic optimization for a target pressure[9]. On top of the research activities, the method has been used for production in Europe as well as in Japan. In Japan, heavy industry companies applied this method to design of a wing for practical aircraft[10].

The above mentioned list of works is not exhaustive. There are likely other notable works involving the inverse design methods of which the first author is not aware.

## Development of the Formulation for Supersonic Flows

A next generation supersonic transport (SST) is of great interest in Japan as well as in Europe and the U.S. because of the projected trends in the world aviation market in the near future. Japan started an SST program in 1995 and will conduct the first flight test of an experimental scaled SST in 2001. The program, which the National Aerospace Laboratory (NAL) organizes, requires advanced CFD technology, especially to determine the aerodynamic shape[13]. One of the most important challenges is the improvement of the L/D ratio during cruising. The NAL aims to design a Natural Laminar Flow (NLF) wing to reduce drag. To realize an NLF wing, the pressure distribution on the upper surface of a wing, which is a function of both the camber curvature and the thickness distribution, is of primary importance. However, so far, supersonic wing design has been performed considering the load distribution on the wing. Accordingly, most existing methods treat only the camber of the wing, because the thickness does not affect the load. We have been developing and verifying a numerical inverse design method for a supersonic wing with Jeong, Iwamiya, Obayashi, and Nakahashi, with the intention of applying it to the design of the Japanese SST[14,15]. The development is the extension of the design method introduced before in this article. The original method can treat both the camber curvature and the thickness distribution simultaneously, but it is for a transonic wing. In this project, the formulation itself has been newly performed for supersonic flows. And, the design of a wing for a wing-fuselage configuration has been conducted. The principals of the new method and the design results are presented in the following three sections.

## New Formulation for Supersonic Flows

The flowfield considered here is shown in Fig.4. A wing is located at $z = 0$ in a supersonic flowfield with freestream Mach number $M_\infty > 1$. The $x$-axis is streamwise, the $y$-axis is spanwise. The free stream velocity vector is assumed as $(1, 0, 0)$. The basic equations are the same as the original formulation except that the flow field is assumed to obey the linearized small disturbance equation.

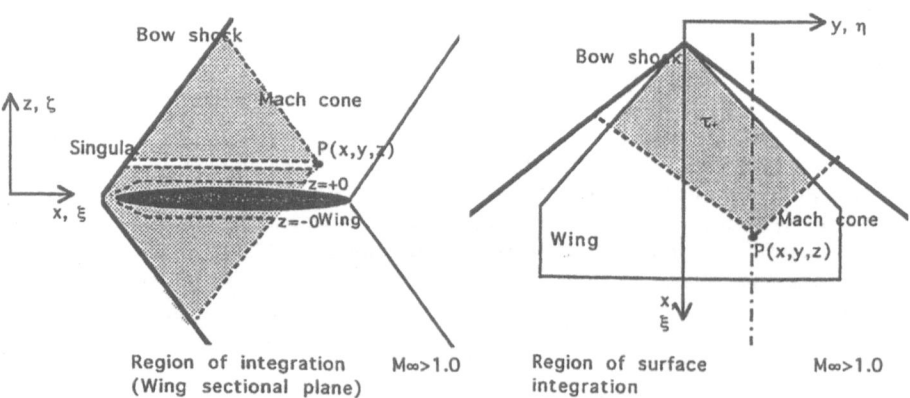

Fig. 4 Flowfield and coordinate system for formulation ($M_\infty > 1.0$).

The formulation starts with the following equations corresponding to Eqs.(7-9), respectively.

$$(1 - M_\infty^2)\Delta\overline{\phi}_{\overline{xx}} + \Delta\overline{\phi}_{\overline{yy}} + \Delta\overline{\phi}_{\overline{zz}} = 0 . \tag{52}$$

The shape of a wing is expressed as $f_\pm(\overline{x}, \overline{y})$. + indicates the upper surface and − indicates the lower surface.

$$\Delta\overline{\phi}_z(\overline{x}, \overline{y}, \pm 0) = \frac{\partial}{\partial\overline{x}}\Delta f_\pm(\overline{x}, \overline{y}) \tag{53}$$

$$\Delta Cp_\pm(\overline{x}, \overline{y}) = -2\Delta\overline{\phi}_x(\overline{x}, \overline{y}, \pm 0) \tag{54}$$

where +0 and -0 mean the upper and lower surfaces of the wing respectively.

Applying a Prandtl-Glauert-like transformation such as

$$x = \overline{x}, \quad y = \beta\overline{y}, \quad z = \beta\overline{z}, \quad \phi(x, y, z) = \frac{1}{\beta^2}\overline{\phi}(\overline{x}, \overline{y}, \overline{z})$$

where $\beta = \sqrt{M_\infty^2 - 1}$.

The equations for variation of the perturbation velocity potential, the correction in wing section shapes, and surface pressure difference between one state and another are

$$-\Delta\phi_{xx} + \Delta\phi_{yy} + \Delta\phi_{zz} = 0 \tag{55}$$

$$\frac{\partial}{\partial x}\Delta f_\pm(x, \frac{y}{\beta}) = \beta^3 \Delta\phi_z(x, y, \pm 0) \tag{56}$$

$$\Delta Cp_\pm(x, \frac{y}{\beta}) = -2\beta^2 \Delta\phi_x(x, y, \pm 0) . \tag{57}$$

Now, we apply Green's theorem to Eq.(54) of a hyperbolic system and obtain $\Delta\phi$ in an analytical form,

$$\Delta\phi(x, y, z) = -\frac{1}{2\pi}\frac{\partial}{\partial x}\int\int_{\tau_+}\left\{[\Delta\phi_\zeta(\xi, \eta, +0) - \Delta\phi_\zeta(\xi, \eta, -0)]\Psi(x, y, z; \xi, \eta, 0)\right\}d\xi d\eta$$

$$+\frac{1}{2\pi}\frac{\partial}{\partial x}\int\int_{\tau_+}\left\{[\Delta\phi(\xi, \eta, +0) - \Delta\phi(\xi, \eta, -0)]\Psi_\zeta(x, y, z; \xi, \eta, 0)\right\}d\xi d\eta \tag{58}$$

where

$$\Psi(x, y, z; \xi, \eta, \zeta) = \cosh^{-1}\frac{x - \xi}{\sqrt{(y - \eta)^2 + (z - \zeta)^2}} \tag{59}$$

$$\Psi_\zeta(x, y, z; \xi, \eta, \zeta) = \frac{(x - \xi)(z - \zeta)}{[(y - \eta)^2 + (z - \zeta)^2]}\frac{1}{\sqrt{(x - \xi)^2 - (y - \eta)^2 - (z - \zeta)^2}} . \tag{60}$$

Then, the formulation is performed referring to the formulation in the forth section which is for transonic wing ($M_\infty < 1.0$) design. Unlike a transonic flowfield which is mainly an elliptic system, the domain influenced by the disturbance at $P(x, y, z)$ is limited. So the domain for integrations ,which is a portion of the physical space around a wing, should be carefully defined for Eq.(58). Since the influenced domain is behind the bow shock wave and inside the Mach forecone from $P(x, y, z)$, the integral domain is bounded with the two surfaces in the manner shown in the first figure of Fig.4. On the surface of the bow shock wave and the Mach cone, each integrand in Eq.(58) becomes zero. The surface integral domain where the integration remains nonzero is a portion of the wing surface which is bounded by the leading edge line and the hyperbola $(x - \xi)^2 - (y - \eta)^2 - (z)^2 = 0$. In Eq.(58), every integrand is divided into two functions; one for the upper surface($\zeta = +0$) and the other for the lower surface($\zeta = -0$). So, the domain $\tau_+$ means the upper surface of the wing plane where $(x - \xi)^2 - (y - \eta)^2 - (z)^2 \geq 0$. It is shown as a shadowed area in Fig.4.

In order to expose the boundary condition $\Delta Cp$ and the unknown shape function $\Delta f$ as explicit functions, we do further calculus with Eq.(58). In fact, $\Delta Cp$ is associated with $\Delta \phi_x$ and $\Delta f$ is associated with $\Delta \phi_z$. Differentiating Eq.(58) with respect to $x$ and adding $\Delta \phi_x(x, y, z)$ at $z = +0$ to $\Delta \phi_x(x, y, z)$ at $z = -0$, we obtain

$$\Delta w_s(x, y) = -\Delta u_s(x, y) - \frac{1}{\pi} \int \int_{\tau_+} \frac{(x - \xi)\Delta w_s(\xi, \eta)}{[(x - \xi)^2 - (y - \eta)^2]^{3/2}} d\xi d\eta \qquad (61)$$

where

$$\Delta u_s = \Delta \phi_x(x, y, +0) + \Delta \phi_x(x, y, -0) = -\frac{1}{2\beta^2}(\Delta Cp(x, \frac{y}{\beta}, +0) + \Delta Cp(x, \frac{y}{\beta}, -0)) \qquad (62)$$

$$\Delta w_s = \Delta \phi_z(x, y, +0) - \Delta \phi_z(x, y, -0) = -\frac{1}{\beta^3}(\frac{\partial \Delta f(x, \frac{y}{\beta}, +0)}{\partial x} - \frac{\partial \Delta f(x, \frac{y}{\beta}, -0)}{\partial x}). \qquad (63)$$

Similarly, differentiating Eq.(58) with respect to $z$ and adding $\Delta \phi_z(x, y, z)$ at $z = +0$ to $\Delta \phi_z(x, y, z)$ at $z = -0$

$$\Delta w_a(x, y) = -\Delta u_a(x, y) + \frac{1}{\pi} \int \int_{\tau_+} \frac{(x - \xi)\Delta u_a(\xi, \eta)}{(y - \eta)^2 \sqrt{(x - \xi)^2 - (y - \eta)^2}} d\xi d\eta \qquad (64)$$

where

$$\Delta u_a = \Delta \phi_x(x, y, +0) - \Delta \phi_x(x, y, -0) = -\frac{1}{2\beta^2}(\Delta Cp(x, \frac{y}{\beta}, +0) - \Delta Cp(x, \frac{y}{\beta}, -0)) \qquad (65)$$

$$\Delta w_a = \Delta \phi_z(x, y, +0) + \Delta \phi_z(x, y, -0) = -\frac{1}{\beta^3}(\frac{\partial \Delta f(x, \frac{y}{\beta}, +0)}{\partial x} + \frac{\partial \Delta f(x, \frac{y}{\beta}, -0)}{\partial x}). \qquad (66)$$

The same fundamental equations for pressure and surface geometry are found in Ref.[16]. Eq.(61) is a Volterra integral equation of the second kind for $\Delta w_s$, the thickness change at $(x, y)$ on a wing, while Eq.(64) is the integral expression for $\Delta w_a$, the curvature change of the wing section camber, at $(x, y)$. There needs to be special treatment for the integration, because the integrands of Eqs.(61) and (64) become singular on the Mach cone.

To guarantee that every section has a closed trailing edge, the solution $\Delta w_s$ is modified as

$$\Delta w_s^{mod}(x, y) = \Delta w_s(x, y) - \frac{\int_{L.E.}^{T.E.} \Delta w_s(\xi, y) d\xi}{\int_{L.E.}^{T.E.} d\xi} \qquad (67)$$

so as to satisfy the condition:

$$\int_{L.E.}^{T.E.} \Delta w_s^{mod}(\xi, y)d\xi = 0 .$$

(68)

The geometrical correction is calculated using

$$\Delta f_\pm(x, \frac{y}{\beta}) = \frac{1}{2}\beta^3 \int_{L.E}^x \left[\Delta w_s^{mod}(\xi, y) \pm \Delta w_a(\xi, y)\right]d\xi .$$

(69)

Therefore, we can obtain the geometrical correction everywhere on a wing, specifying the difference between target and current pressures, $\Delta Cp = Cp^{target} - Cp^{current}$.

## Wing Design for Wing-Fuselage Combination

To develop a practical airplane, the wing-fuselage combination should be handled. One cannot design a high performance wing without taking the wing-fuselage interaction into consideration. For the aerodynamic design of an SST wing, the method with the supersonic inverse problem has been applied to the wing-fuselage configuration. Originally, the design method is developed for a wing alone. Then, we have to make two modifications in the design loop. First, simulation about a wing-fuselage configuration is conducted in the analysis part[17]. Second, concerning the input data for the inverse design part, extra data are added to the data of the wing geometry and pressure directly obtained through the analysis part. To solve the inverse problem, we should not use the pressure and geometry data of the fuselage part, because the formulation has been done with the thin wing theory. When analyzing the fowfield, the wing ends at the root. On the other hand, for solving the inverse problem, it should not end there but continue to the fuselage center. However, no information is given for the part from the root section to the center $(y = 0)$ of the wing, because that part is overlapped by the fuselage. So, extra wing section data to represent the hidden part should be prepared in a reasonable way. We generate them by linearly extrapolating geometrical coordinates and pressure distributions of the span section at the root of the wing. With these two modifications, the designed wing geometry reflects the effect of the wing-fuselage interaction when the fuselage remains unchangeed and the variation of the wing-fuselage interaction by the geometrical correction of the wing is small enough. This is owed to the residual-correction concept of the method and the $\Delta$-form formulation of the inverse problem.

## Application to the Japanese SST

The wing of the Japanese SST is aerodynamically designed at $M_\infty = 2.0$. The SST planform is illustrated in Fig.5. To design a high L/D wing, we prescribe a target pressure whose elliptical load distribution minimizes the induced drag and whose upper surface distribution keeps the laminar boundary layer significantly longer than traditional wings. The goal is to design a section geometry for the SST wing which realizes NLF on its upper surface and the optimum load distribution.

For the computation, the half span of the wing is divided into 82 (spanwise) × 50 (chordwise) panels.

The baseline shape of the wing is the result of planform and warp optimizations in terms of the L/D ratio, while the shape of the fuselage is determined using the area rule[13]. The thickness distribution of the NACA66003 airfoil is adopted as the chordwise thickness distribution for each span station of the wing. Despite the optimizations, the performance of the wing of the wing-fuselage model is not as efficient as expected. This is because these optimizations were done for a wing alone. In other words, they did not take the wing-fuselage interaction into consideration. Therefore, improvement of the aerodynamic shape of the SST wing by a method which can account for the interaction is necessary.

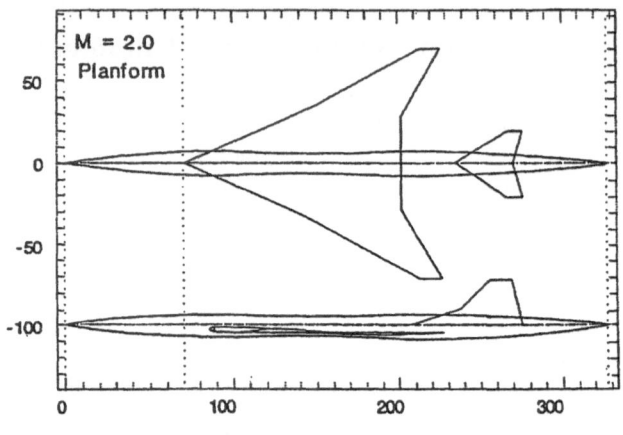

Fig. 5 SST Planform.

The design results at several span stations are shown in Figs.6-10. Fig.6 presents the wing section geometry and the realized pressure distribution along the chord at the 15%-span station. The dashed line and '+-line' indicate, respectively, the geometry and pressure distribution of the baseline wing section, while the solid line and '◇-line' indicate those of the designed wing section. The target pressure is indicated by chain lines. The pressure distibution on the wing surface is calculated by the Navier-Stokes flow solver[17]. Fig.7 shows the wing section geometry and pressure distribution at the 30%-span station. Fig.8 shows those at the 50%-span station. Figs. 9 and 10 show those at the 70% and 90%-span stations, respectively. The resulting wing realizes almost identical pressure distribution to the target. Even at a root section (15%-span station) and in the vicinity of a tip (90%-span station), this method works well, as shown in Figs.6 and 10. One of the most characteristic features of an NLF wing is the sudden expansion of the upper surface $Cp$ distribution at the leading edge. Another one is a flat roof type of $Cp$ distribution along the chord on the upper surface. Those features are considered to make turbulence transition take place as late as possible. At every span-station, the designed wing produces a $Cp$ distribution which has the desired features for an NLF wing.

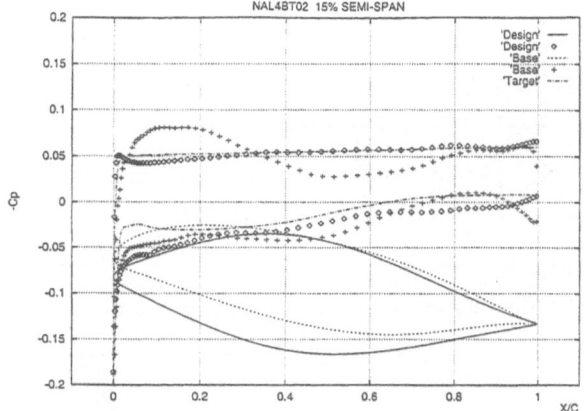

Fig.6 Cp distribution and section geometry at 15% semi-span.

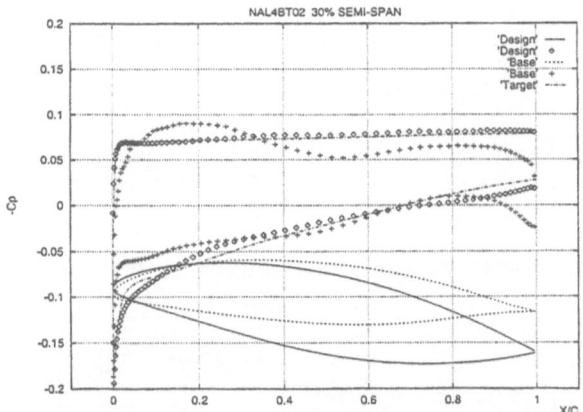

Fig.7 Cp distribution and section geometry at 30% semi-span.

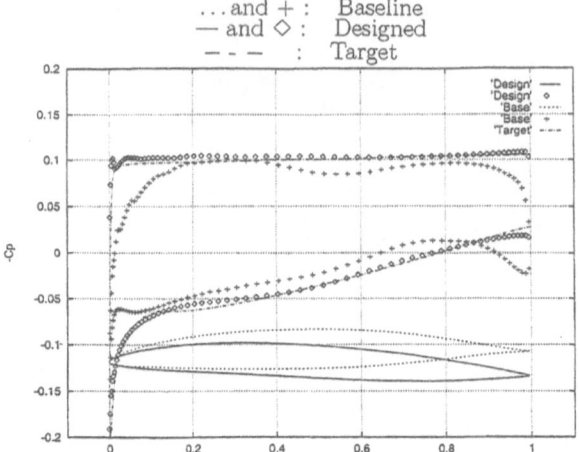

Fig.8 Cp distribution and section geometry at 50% semi-span.

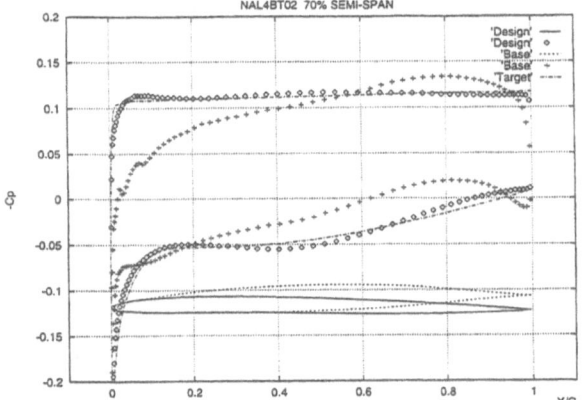

Fig.9 Cp distribution and section geometry at 70% semi-span.

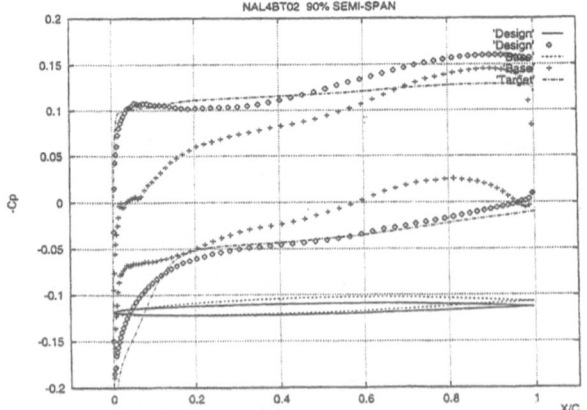

Fig.10 Cp distribution and section geometry at 90% semi-span.

# Development of the Formulation for Multi Objects

As mentioned before, a considerable number of formulations of inverse problems have been reported, so far. Most of those inverse problems aim to design a single object in a flowfield. Difficulty is sometimes experienced when one needs to design both of two wings in a flowfield, which are mutually interacting with each other. Using an inverse problem for a single wing, in order to design the two wings, one has to handle the multi-point design problem. In the case of the multi-point design, it is difficult to prevent the performance of one wing from being disturbed by the improvement of that of the other wing, especially when the interaction between the two wings is strong. It is because the single-wing formulation never takes the mutual interacting effect among wings into consideration. There are several inverse design methods devised for multi-element airfoils[18,19]. They use the corresponding inverse problems. Those methods are effective only for two-dimentional potential flows.

Recently, the authors have extended the original method, pursuing the reliable and efficient design of multiple objects (wings) in a flowfield[20,21]. The inverse problem has been newly formulated for multiple wings in a flowfield. The residual-correction design method with the new inverse problem can also treat three-dimentional problems and viscous (Navier-Stokes) flows, like the original one. The new formulation has brought several new terms to account for interaction between wings. The new terms make the integral equation system of an inverse problem much more complicated. Nevertheless, it can be solved without a huge amount of computational cost. That interaction is never evaluated if the design is performed by the single-wing inverse problem.

The new method is useful to design a multi-element wing for high-lift, dual wing systems and tandem wing systems. It provides the geometry for all of the objects in a flowfield simultaneously.

# New Formulation for Multiple Wing Systems

In this section, an integral equation system for an inverse problem is derived in order to determine shapes of multiple wings in a flowfield simultaneously. A flowfield where $kmax$ wings, Wing-1, Wing-2...and Wing-$kmax$, exist is considered. The flowfield with $kmax (= 2)$ wings and its coordinates system for the formulation are indicated in Fig.11. The $x$ axis is streamwise, the $y$ axis spanwise and the $z$ axis is in the thickness direction of the wings. The free stream velocity vector is assumed to be $(1, 0, 0)$. The free stream Mach number is $M_\infty$ and $\gamma$ is the ratio of specific heats.

The concept of the formulation is, again, to build a mathematical model which relates a geometrical correction, $\Delta f$, to the pressure difference, $\Delta Cp$. The formulation starts with $\Delta$-form equations such as Eqs.(10-12) in the third section. In the plural wing case, we have a flow equation for the velocity potential, $kmax$ boundary geometry equations and $kmax$ boundary surface pressure equations;

$$\Delta\phi_{xx} + \Delta\phi_{yy} + \Delta\phi_{zz} = \frac{\partial}{\partial x}\Big[\frac{1}{2}(\phi_x + \Delta\phi_x)^2 - \frac{1}{2}\phi_x^2\Big] \qquad (70)$$

$$\Delta\phi_z(x, y, c_k \pm 0) = \frac{\partial}{\partial x}\Delta f_{k\pm}(x, y) \quad (k = 1, 2, ...kmax) \tag{71}$$

$$2\Delta\phi_x(x, y, c_k \pm 0) = -\frac{K}{2\beta^2}\Delta Cp_{k\pm}(x, \frac{y}{\beta}) \quad (k = 1, 2, ...kmax) . \tag{72}$$

The shape of Wing-$k$ is expressed as $f_{k\pm}(x, y)$ where $+0$ and the subscript $+$ stand for the upper surface and $-0$ and the subscript $-$ indicate the lower surface. $c_k$ is the $z$ location of the mean plane of Wing-$k$.

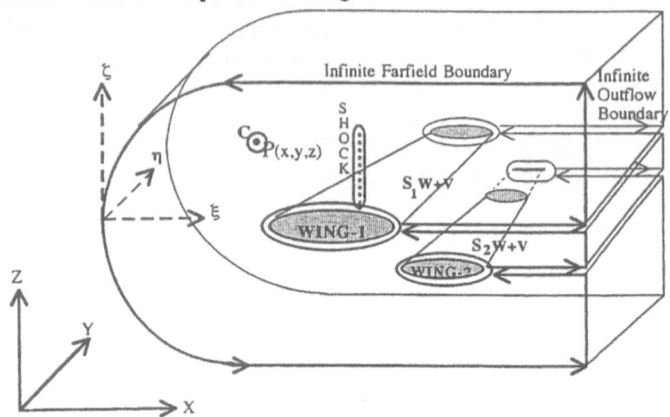

Fig.11 Flowfield and coordinate system.

Incidently, one of the most common and practical designs of multiple wing systems is that for multi-element wings of high-lift. When the design of a high lift system is intended, $M_\infty$ of the design problem is low. When the flow is low subsonic, Eq.(69) yields

$$\Delta\phi_{xx} + \Delta\phi_{yy} + \Delta\phi_{zz} \simeq 0 . \tag{73}$$

To make the discussion compact, the formulation is performed using Eq.(73) here. This is also consistent with the design problem shown in the later section. There, we demonstrate the method of the new inverse formulation on a high-lift system of a main wing and a flap in a low subsonic flowfield. Interested readers would find the formulation without the omission the right-hand-side of Eq.(70), as well as the design problems in transonic flowfields, in Ref.[20] and [21].

Applying Green's theorem to Eq.(73), and then performing some calculus as shown in the forth section, two integral equations for each wing, *i.e.* $2kmax$ integral equations in total, are obtained.

$$\Delta U s_k(x, y) = -\frac{1}{2\pi}\int\int_{s_k w}\Big[\Psi_x(x, y, c_k; \xi, \eta, c_k)\Delta W s_k(\xi, \eta)\Big]d\xi d\eta$$

$$-\frac{1}{2\pi}\sum_{p\neq k}\int\int_{s_k w}\Big[\Psi_x(x, y, c_k; \xi, \eta, c_p)\Delta W s_p(\xi, \eta)\Big]d\xi d\eta$$

$$+\frac{1}{2\pi}\sum_{p\neq k}\int\int_{s_k w}\Big[\Psi_\zeta(x, y, c_k; \xi, \eta, c_p)\Delta U a_p(\xi, \eta)\Big]d\xi d\eta$$

$$(k = 1, 2, ...kmax) \tag{74}$$

$$\Delta W a_k(x,y) = \frac{1}{2\pi} \int \int_{s_{kw}} \left\{ \frac{\Delta U a_k(\xi,\eta)}{(y-\eta)^2} \left[ 1 + \frac{x-\xi}{\sqrt{(x-\xi)^2 + (y-\eta)^2}} \right] \right\} d\xi d\eta$$

$$+ \frac{1}{2\pi} \sum_{p \neq k} \int \int_{s_{pw}} \left\{ \frac{\Delta U a_p(\xi,\eta)}{(y-\eta)^2 + \bar{c}_{k,p}^2} \left[ 1 + \frac{x-\xi}{\sqrt{(x-\xi)^2 + (y-\eta)^2 + \bar{c}_{k,p}^2}} \right] \right\} d\xi d\eta$$

$$- \frac{1}{2\pi} \sum_{p \neq k} \int \int_{s_{pw}} \left[ \Psi_z(x,y,c_k; \xi, \eta, c_p) \Delta W s_p(\xi, \eta) \right] d\xi d\eta$$

$$- \frac{1}{2\pi} \sum_{p \neq k} \int \int_{s_{pw}} \Delta U a_p(\xi, \eta) \frac{\bar{c}_{k,p}^2}{[(y-\eta)^2 + \bar{c}_{k,p}^2]^2} (2 + 3q - q^2) d\xi d\eta$$

$$(k = 1, 2, ...kmax) \tag{75}$$

where $\quad \Delta W s_k(x,y) = \Delta \phi_z(x,y,c_k+0) - \Delta \phi_z(x,y,c_k-0)$

$$\Delta W a_k(x,y) = \Delta \phi_z(x,y,c_k+0) + \Delta \phi_z(x,y,c_k-0)$$

$$\Delta U s_k(x,y) = \Delta \phi_x(x,y,c_k+0) - \Delta \phi_x(x,y,c_k-0)$$

$$\Delta U a_k(x,y) = \Delta \phi_x(x,y,c_k+0) + \Delta \phi_x(x,y,c_k-0)$$

$$\Psi(x,y,z; \xi, \eta, \zeta) = \frac{1}{\sqrt{(x-\xi)^2 + (y-\eta)^2 + (z-\zeta)^2}}$$

$$q = \frac{x-\xi}{\sqrt{(x-\xi)^2 + (y-\eta)^2 + \bar{c}_{k,p}^2}}, \qquad \bar{c}_{k,p} = c_k - c_p .$$

$\Delta U s_k$ and $\Delta U a_k$ are calculated from the pressure difference, $\Delta C p_{k\pm}$. $\Delta W s_k$ and $\Delta W a_k$ are the unknowns for the equation system. They are interpreted into the geometry correction function of each Wing-k ($k = 1, 2, ...kmax$). $\Delta W s_k(x,y)$ is associated with the thickness change at $(x,y)$ on wing-k, while $\Delta W a_k(x,y)$ is the curvature change of the camber of Wing-k. Eqs.(74) and (75) correspond Eqs.(31) and (32), respectively. The second terms of both of Eqs.(74) and (75) represent the linearly accumulating effect of interaction by the other wings in the flowfield. The last term of the surface integral in Eq.(74) and the third and forth terms of Eq.(75) are the mathematical functions representing the nonlinear effect among wings.

To guarantee that every section has a closed trailing edge, the solution $\Delta W s$ is constrained with the condition:

$$\int_{L.E.}^{T.E.} \Delta W s_k(\xi, y) d\xi = 0 \quad (k = 1, 2, ...kmax) . \tag{76}$$

The geometrical correction is calculated using

$$\Delta f_{k\pm}(x,y) = \int_{L.E}^{x} \left[ \Delta W s_k(\xi, y) \pm \Delta W a_k(\xi, y) \right] d\xi \quad (k = 1, 2, ...kmax) . \tag{77}$$

Therefore, the geometrical correction of every wing is obtained in an analytical function form on each wing surface. Then, each wing surface is divided into small panels as shown in Fig.12. Eqs.(74)and(75) yield algebraic equations which can be solved by numerical computation.

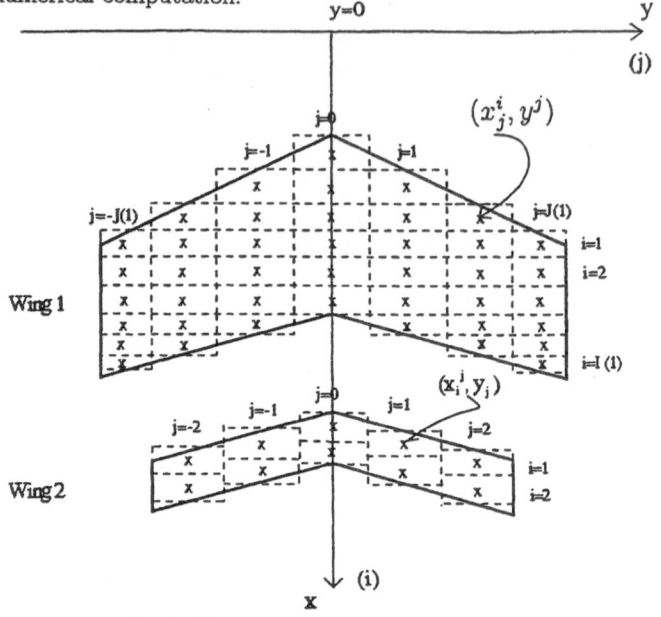

Fig.12 Discretization of wing surfaces.

## Design of a Multi-element Wing

A two element wing, which consists of a main part and a flap, has been designed in a subsonic flow of $M_\infty = 0.2$. For conciseness, the wing is assumed to be a infinite-long span rectangular wing, because this is the preliminary trial to validate the feasibility of the method to design high-lift systems. The initial geometry of wing sections at the mid-span ($y = 0$) is shown in Fig.13. The corresponding initial pressure distribution is also shown in Fig.13. The pressure distrubution is calculated using a Navier-Stokes flow solver[22]. Every section of the baseline wing is an NACA0012 airfoil with 0 degree angle of attack. The chord length of the main part is 1.0 while that of the flap is 0.35. The $x$ and $z$ distances between the two objects, $\Delta x$ and $\Delta z$, are 0.1 and 0.05, respectively. $M_\infty$ is 0.20 and the Reynolds number is $1.0x10^6$. The target pressure distribution is specified as indicated by solid lines in Fig.14.

Fig.14 shows the history of the design process. The current pressure distribution and geometry at the mid-span section at y=0 of both of the main part and the flap are presented. The specified target pressure distribution is also presented for purpose comparison. The two sub-figures lined on the top row of Fig.14 show the situation at the initial stage. The initial pressure distribution is plotted with symbols, '+' indicating the upper surface pressure and 'x' the lower one. The lift

Multi-Element Wing

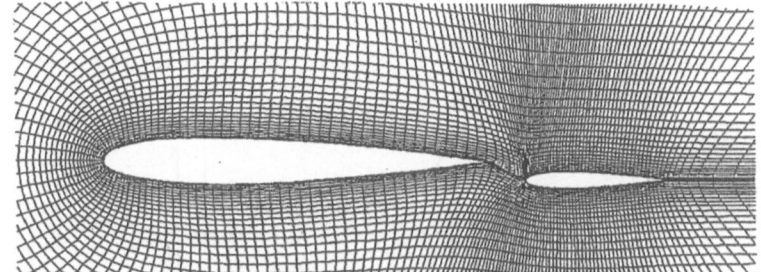

| Front | : | NACA0012 | Chord=1.0 | $\alpha$=0° |
|-------|---|----------|-----------|------|
| Rear  | : | NACA0012 | $''$ =0.35 | $''$ =0° |
| Distance | : | $\Delta x$ = 0.1 | $\Delta z$ = 0.05 | |

Initial pressure distribution

Flow condition :    $M_\infty$= 0.20    Re = 1.0 million

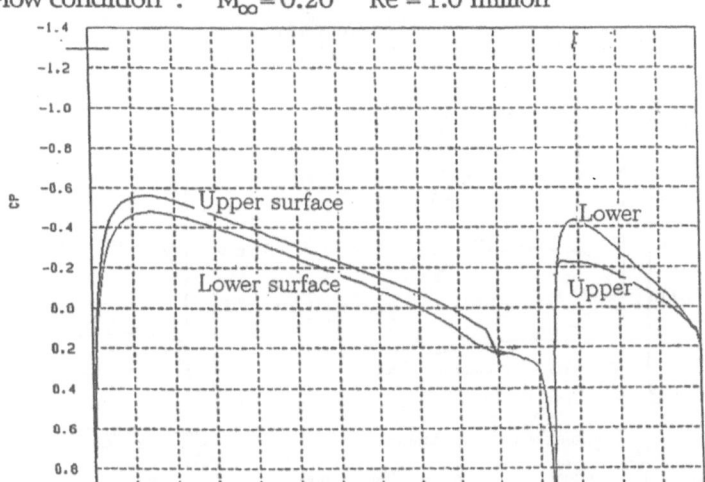

Fig.13 Section geometry and Cp distribution at the initial stage.

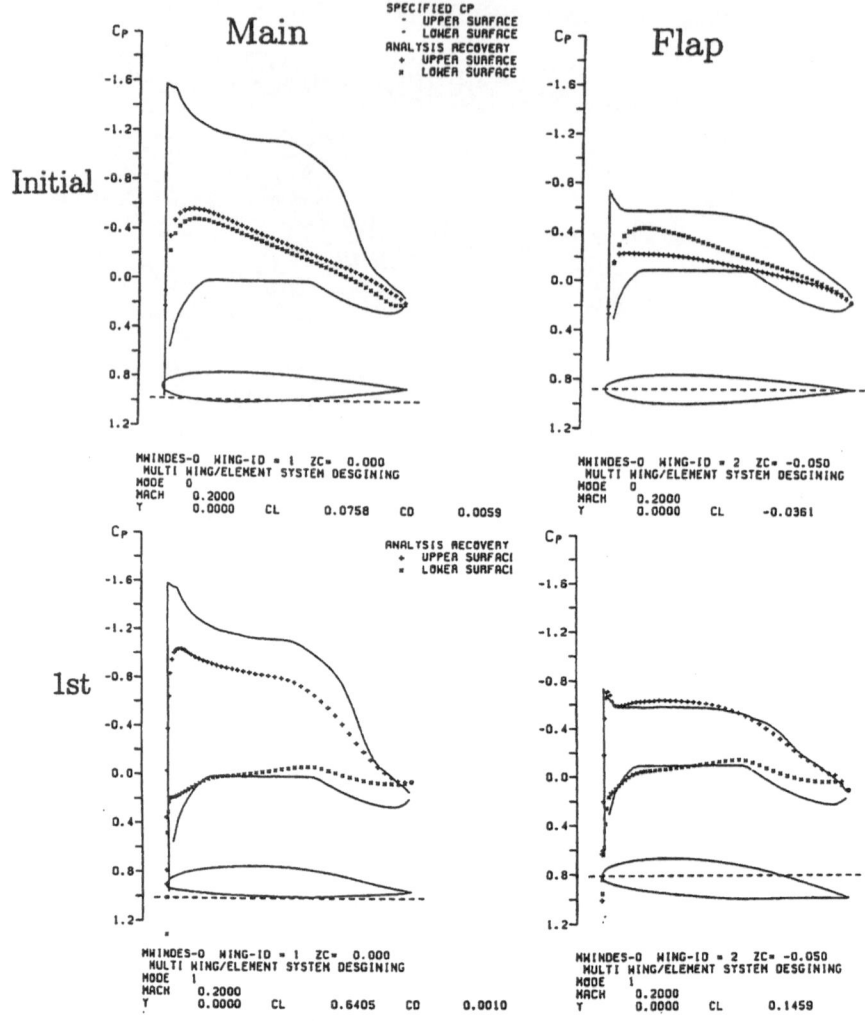

Fig.14 History of design process, Mach=0.20, Re=1million,
Target Cp: solid lines,     Current Cp: + and ×
(to be continued to the next page).

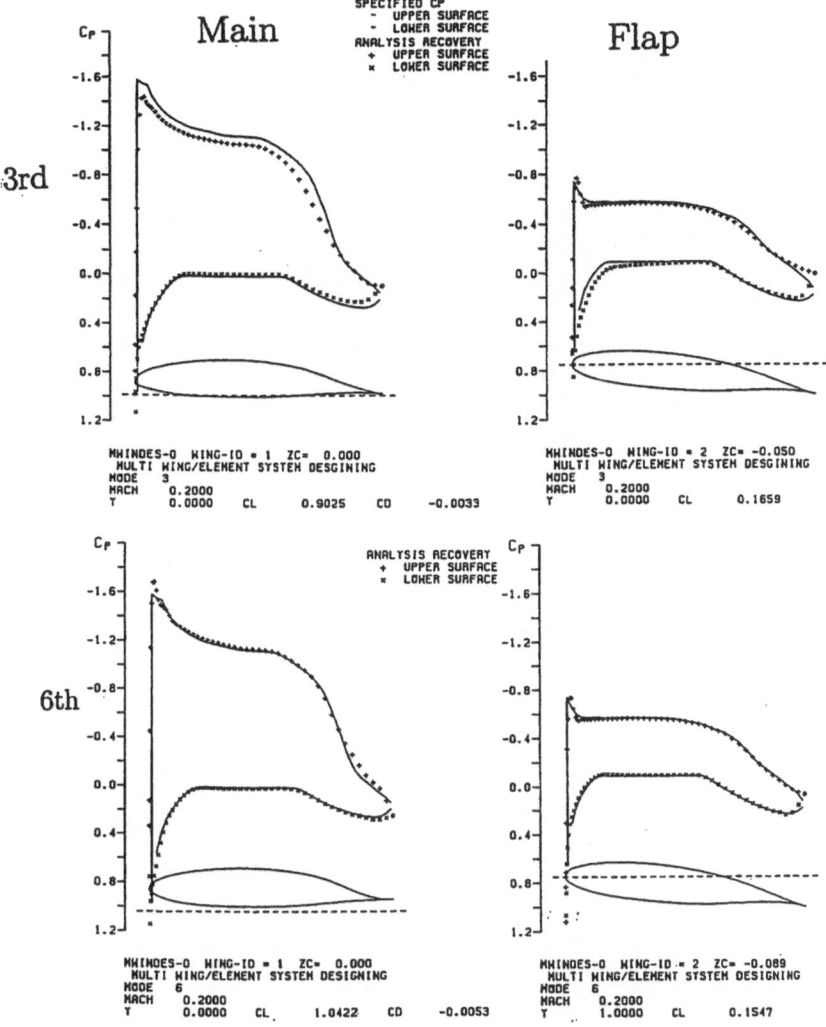

Fig.14 History of design process, Mach=0.20, Re=1million,

Target Cp: solid lines,     Current Cp: + and ×

<u>Example</u>   Multi-Element Wing

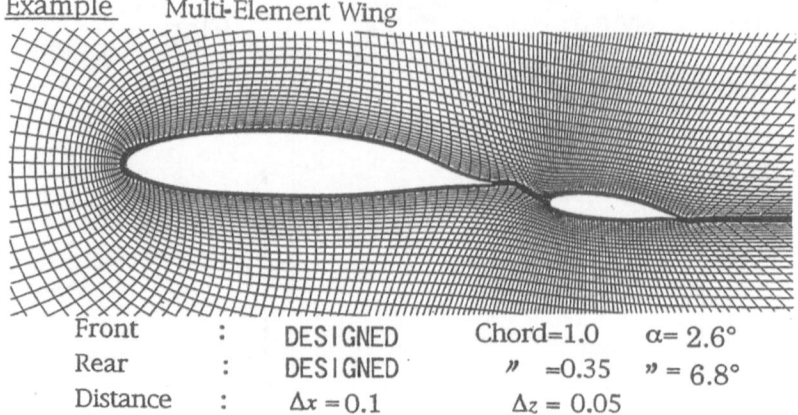

| Front | : | DESIGNED | Chord=1.0 | α= 2.6° |
|---|---|---|---|---|
| Rear | : | DESIGNED | " =0.35 | " = 6.8° |
| Distance | : | Δx = 0.1 | Δz = 0.05 | |

Designed pressure distribution
Flow condition :   $M_\infty = 0.20$   Re = 1.0 million

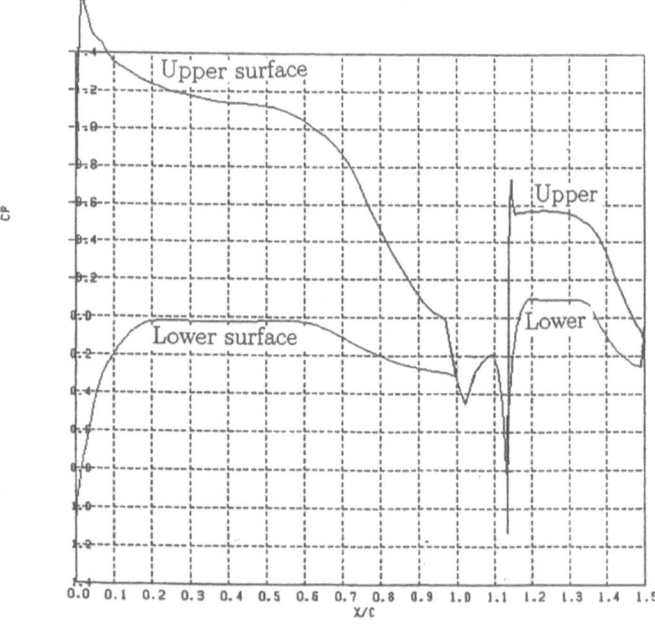

Fig.15 Section geometry and Cp distribution at the converged stage.

of the main part is slightly positive and that of the flap is negative. This is also seen in Fig.13. These facts imply the interaction effect between the two parts. If there were no interaction, the lift of each part would be zero because both the main part and the flap have symmetrical shapes along the chord and zero angle of attack. The other sub-figures in Fig.15 indicate the situations after one iteration of the design loop, after three iterations, and after six iterations, respectively. In each sub-figure, the current pressure distribution is plotted with symbols, '+' and 'x'. After six iterations, the design can be regarded to be complete. Fig.15 shows the shape of the designed wing sections of the main part and the flap in addition to the correspoding pressure distribution. The comparison of the initial and designed geometry is made in Fig.16. The airfoils are drawn to five times scale in the $z$ direction. Both parts of the designed wing have become thicker. The main part has come to take 2.6° angle of attack, and the flap has come to take 6.8° angle of attack.

The method has worked well on the design of a two-element airfoil of high lift systems. It has shown that the method can achieve complicated aerodynamic design of multi-element wings. For this example, the required computational time to perform the one iteration loop, consisting of the Navier-Stokes flow analysis[22] and the inverse problem, is about 30 minutes on a 1.7 GFLOP computer.

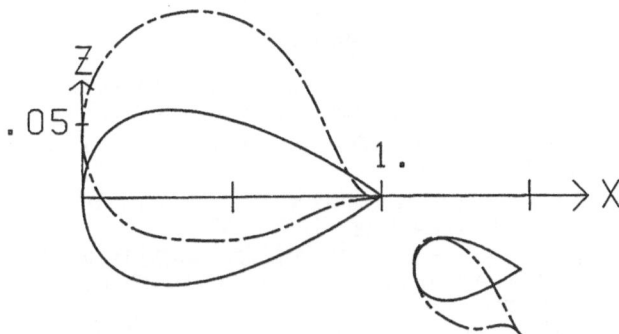

Fig. 16 Comparison of initial and designed shapes
Initial: solid lines,    Designed: chain lines.

## Summary

An aerodynamic design method was reviewed and studied. Two other recently developed design methods were also presented. The first was devised for the design of transonic wings. It was explained that the methodology was chosen to make the method efficient and versatile. The two new methods derived from the original inverse design method have been implemented for their intended design categories; specifically supersonic wings and multi-element wings. The important factors of the methodology are the residual-correction concept, the $\Delta$-form formulation of an inverse problem, and the inverse problem to solve integral equations describing the physics of flowfields.

For each of the three methods, the corresponding inverse problem was formulated. Integral equations were derived using the small disturbance approximation, the thin wing theory and Green's theorem.

With the residual-correction concept, the inverse problems can be incorporated into any kind of flow analysis. The accuracy of the design methods primarily depends on the analysis part. Then, realistic design problems for Navier-Stokes flows are solvable by the methods.

The original method has a fifteen-year history. During those years, it has gained popularity because of its efficiency and applicability, and has been used for practical transonic wing design.

The applications of the two new methods were briefly discussed. One worked well for a supersonic wing for wing-fuselage configuration. The other showed that the method should be promising for the design of a multi-element wing for high-lift systems.

With any of the three methods, the design results can be obtained within about ten iterations of the design loop. For ten iterations, it requires eleven of flow simulations and ten times solving the inverse problem. This amount of cost seems to be one of the lowest among the existing design methods. Therefore, the inverse design methods discussed here have proven their applicability and efficiency.

## References

[1] Laburujere, Th. E. and Slooff, J. W.: Computational Methods for the Aerodynamic Design of Aircraft Components, *Annu. Rev. Fluid Mech.*, 25 (1993), pp. 183-214.

[2] Dulikravich, G. S.: Shape Inverse Design and Optimization for Three-Dimensional Aerodynamics, *AIAA-95-0695*, 1995.

[3] Takanashi, S.: Iterative Three-Dimensional Transonic Wing Design Using Integral Equations, *J. Aircraft*, Vol. 22, No. 8, pp. 655-660, 1985.

[4] Hirose, N., Takanashi, S. and Kawai, N.: Transonic Airfoil Design Based on Navier-Stokes Equation to Attain Arbitrarily Specified Pressure Distribution - an Iterative Procedure, *AIAA-85-1592*, 1985.

[5] Xia, Z. X., Zhu, Z. Q. and Vu, L. Y.: A Computational Method for Inverse Design of Transonic Airfoil and Wing, *AIAA-93-3482-CP*, 1993.

[6] Lorentzen, L.; Development of Inverse Airfoil Design for Transonic Applications, *FFA TN 1993-37*, 1993.

[7] Hua, J., Yang, Q. Z., Xi, D. K., Zhang Z. Y., Fu, D. W., Zhang Z. L. and Wang L.: Design and Experimental Investigation of Transonic Natural Laminar Flow Wings, *ICAS-94-4,7,3*, 1994.

[9] Obayashi, S. and Takanashi, S.: Genetic Optimization of Target Pressure Distributions for Inverse Design Methods, *AIAA-95-1649-CP*, 1995.

[8] Bartelheimer, W.: An Improved Integral Equation Method for the Design of Transonic Airfoils and Wings, *AIAA-95-1688-CP*, 1995.

[10] Fujii, K. and Takanashi, S.: Aerodynamic Aircraft Design Methods and Their Notable Applications, *ICIDES-III*, pp.31-45, 1991, and References Therein .

[11] Heaslet, M. A. and Spreiter, J. R.: Three Dimensional Transonic Flow Theory Applied to Slender Wings and Bodies, *NACA Report 1318*, 1957.

[12] Nørstrud, H.,: High Speed Flow Past Wings, *NASA CR-2246*, 1973.

[13] Sinbo, Y., Yoshida, K., Iwamiya T., Takaki R., and Matsushima, K.: Aerodynamic Design of Scaled Supersonic Experimental Airplane, *NAL International CFD Workshop on SST Design* , March, 1998.

[14] Jeong, S., Matsushima, K., Iwamiya, T., Obayashi, S., and Nakahashi, K.: Inverse Design Method for Wings of Supersonic Transport. *AIAA-98-0602*, Jan., 1998.

[15] Matsushima, K., Iwamiya, T., Jeong, S. and Obayashi, S.: Aerodynamic Wing Design for NAL's SST Using an Iterative Inverse Approach, *NAL International CFD Workshop on SST Design* , March, 1998.

[16] Lomax, H., Heaslet, M. A. and Fuller, F. B.: Integrals and Integral equations in Linearized Wing Theory, *NACA Rep.* 1054, 1951.

[17] Takaki, R., Iwamiya, T., Aoki, A.: CFD Analysis Applied to the Supersonic Research Airplane, *NAL International CFD Workshop on SST Design* , March, 1998.

[18] Narramore, J. C., and Beaty, T. D.: An Inverse Method for Multielement High-Lift Systems, *AIAA-75-879*, 1975.

[19] Shigemi, M.: A Solution of Inverse Problems for Multi-Element Aerofoils through Application of Panel Method, *Trans. Japan Soc. Aero. Space Sciences*, Vol. 28, No. 80, 1985.

[20] Matsushima, K. and Takanashi, S.: An Inverse Design Method for Transonic Multiple Wing Systems on Integral Equations, *AIAA-96-2465*, June, 1996.

[21] Matsushima, K., Takanashi, S. and Iwamiya, T.: An Inverse Design Method for Transonic Multiple Wing Systems using Integral Equations, J. Aircraft, Vol. 34, No. 3, 1997.

[22] Fujii, K. and Obayashi, S. : High Resolution Upwind Scheme for Vortical Flow Simulations, J. Aircraft, vol. 26, No. 12, pp. 1123-1129, 1989.

# Application of Transonic Inverse Method to the Development of Aerospace Products

Takeshi Kaiden

Nagoya Aerospace Systems Works
Mitsubishi Heavy Industries, Ltd.
10, Oye-cho, Minato-ku, Nagoya, JAPAN
e-mail:takeshi_kaiden@mx.nasw.mhi.co.jp

## Summary

A three-dimensional transonic wing design method which was developed by Dr. S. Takanashi(Ref.1) has been used to various aero-design fields in Mitsubishi Heavy Industries, Ltd.

The inverse code is originally based on a small perturbation equation. The present method has an iterative "residual-correction" concept combined with existing three-dimensional transonic flow analysis code. The main advantages of this approach is that the analysis code is treated as a "black box" and, as a result, it can be easily replaced by any other code. Aerodynamic designers are therefore able to adopt any flow analysis code that is considered to be appropriate for their designing purposes. The geometry is corrected iteratively according to the difference between the specified and computed pressure distribution, starting from a baseline geometry.

This paper shows two applicated examples of this inverse problem. One is for swept wing of typical civil transport. The other is for non-planar wing of H-II Orbiting Plane(HOPE) tipfin configuration.

## Computational Method

The computational flow chart is shown in Fig.1. First of all, initial geometry and target pressure distribution are prepared as input. Then flow properties around the wing are calculated by using the analysis code. A design calculation is next carried out after reallocating the wing section because of the difference of the computational topology. The geometry correction which compensates the residual is computed based on the difference between the computed pressure distribution and the target pressure distribution. Then the new geometry is obtained by adding the correction to the basic geometry and next step starts after reallocating the wing section again. This direct-inverse iteration procedure continues until the residual becomes small.

## Applied Examples

### 1. Swept wing for civil transport

As flow analysis code, a three-dimensional transonic full potential code is adopted. Because the viscous effect is not dominant on the design point in this case.

The wing planform is fixed with aspect ratio of 9.4, a quarter chord line swept angle of 21° , taper ratio of 0.25 and trailling edge kink station of 38% semispan. The design target of Mach number is 0.82 and lift coefficient is 0.51. According to the three-dimensional design point, the corresponding two-dimensional design point was set and desirable pressure distribution with weak shock wave was specified.

The three-dimensional pressure distribution was then determined so as to realize the two-dimensional performance effectively in spanwise direction.

The local lift coefficient around wing tip was specified lower than that of midspan in order to prevent unfavorable tip stall, and corresponding pressure distribution was determined to achieve this section lift. On the other hand, the effective aerodynamic sweep angle is increased in the inboard wing region.

The three-dimensional target pressure distributions are shown in Fig.2 together with calculated ones at iteration more 0, 3, and 5. Only five iterations were required before the target pressure distribution is realized as shown in the figure.

Figure 3 compares the pressure distribution obtained by the calculation with the experimental data at neighborhood of the design point.

The agreement is found to be quite good for all stations except the trailing edge where viscous effects are most prominent.

### 2. Tipfin for HOPE

A three-dimensional Navier-Stokes code is applied in this case. HOPE configuration with a huge tipfin has transonic non-linear characteristics which are caused by complicated phenomena. Navier-Stokes computational result shows that the tipfin induces shock induced separation, no iso-bar pattern and heavy cross flow affecting the flow even on the body surface. These complicated phenomena is not caught by any other flow analysis code.

The tipfin is away from $\eta$ (along the surface)=0.55 of semispan and the cant angle is 15° . The sectional pressure distribution is determined by paying attention to following points.

(1) The suction peak which occurs on the leading edge of the tipfin and outer wing is suppressed.
(2) The shock wave which appears on the upper surface is eliminated.
(3) Isobar design concept is adopted.

Finally the certain transonic supercritical pressure distribution is chosen as the target pressure.

The comparison of surface streamlines is shown in Fig. 4 and 5. For initial configuration, it is clear that the cross flow appears not only on the outer wing but also on the body. After five direct-inverse iterations, the cross flow becomes weaker on the outer wing. Also, concerning the body surface, the cross flow is eliminated perfectly.

Around the tipfin, the strong shock wave and the induced separation which are created for the initial geometry disappears for the final geometry.

## Conclusion

Takanashi's inverse method has been applied to the aero design of aerospace products in Mitsubishi Heavy Industries, Ltd. As shown here, it is confirmed that this method is a powerful and useful tool.

As future plans, this method will be applied to other various aerospace products.

## References

[1]Takanashi,S.,"Iterative Three-Dimensional Transonic Wing Design Using Integral Equations," Journal of Aircraft Vol. 22, No. 8, Aug. 1985.

[2]Tatsumi,S. and Takanashi,S., "Experimental Verification of Three-Dimensional Transonic Inverse Method," AIAA-85-4077, Oct. 1985.

[3]Kaiden,T., Ogino,J. and Takanashi,S.,"Non-planar Wing Design by Navier-Stokes Inverse Computation," AIAA-92-0285, Jan. 1992.

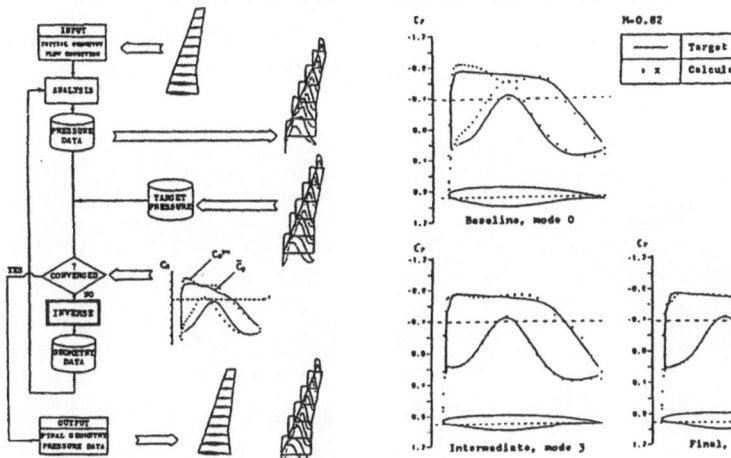

Fig.1  Schematics of three-dimensional inverse method [Ref.2]

Fig.2  Improvement of pressure distribution at 30% semispan station [Ref.2]

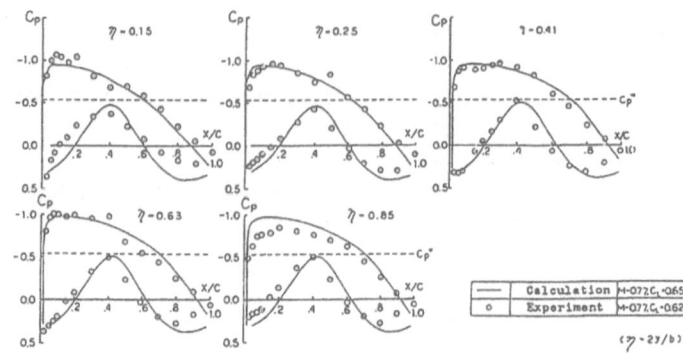

Fig.3　Comparison of calculated wing pressure with experiment　[Ref.2]

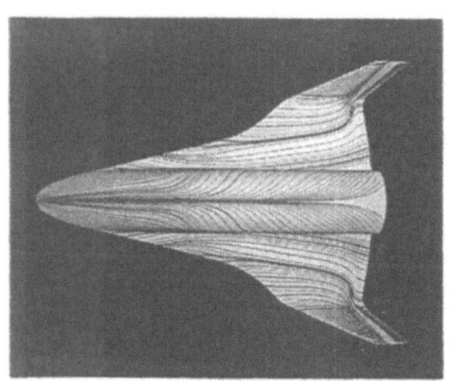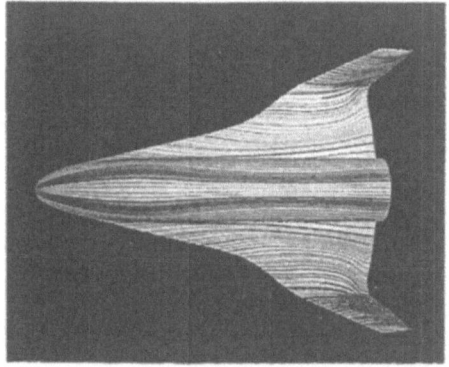

For initial geometry　　　　　　　　For final geometry
Fig.4　Surface streamlines (M=0.9 , $\alpha$ =5° )　[Ref.3]

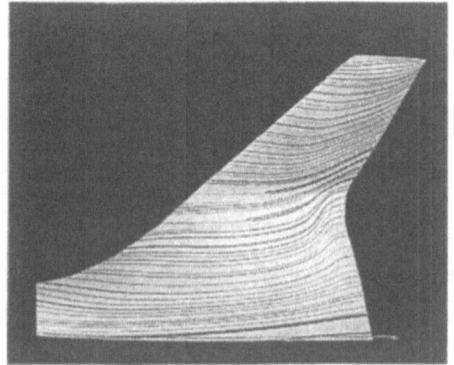

For initial geometry　　　　　　　　For final geometry
Fig.5　Surface streamlines (Tipfin upper side : M=0.9 , $\alpha$ =5° )　[Ref.3]

# Aerodynamic Design of Wing-Engine Configuration under Effect of Jet Plume

Hitoshi Takahashi*, Susumu Takanashi*(f),
Takuji Kishimoto**, Kenji Hayama** and Eiji Shima**

*National Aerospace Laboratory
7-44-1, Jindaiji-Higashi-Machi Chofu City, Tokyo, 182-8522, JAPAN
**Kawasaki Heavy Industries LTD.
1, Kawasaki-Cho, Kakamigahara City, Gifu-Pref. 504-8710, JAPAN

## Summary

The wing-engine configuration is aerodynamically designed by the WINDES, inverse design code developed by Takanashi combined with the multi-block Navier-Stokes solver. The jet plume is considered in the analysis mode and the effect of the jet on/off is discussed. The inverse design method used here improves the L/D performance. Although the target $M_{DD}$ is not achieved, the $M_{DD}$ is also improved.

## Introduction

The powered high lift system using USB(Upper Surface Blowing), in which engines are mounted on wing, is favorable for aerodynamic design of quiet STOL(Short Take-Off and Landing) airplane, since the jet exhaust is covered by the wing and the noise to the ground is shielded in addition to its high lift coefficient.

USB system, however, causes a serious problem on high speed performance, because engines and jet plume on the wing affect the pressure distribution on upper surface of the wing which is most important for aerodynamic performance. In order to improve that, it is necessary to design the wing considering the effect of engine nacelles and jet plume.

Takanashi developed the inverse (design) method in which original shape is modified based on the panel method using the difference of actual and target pressure distribution. After all, this method makes up the shape whose pressure distribution is prescribed by a user, if the pressure distribution is realizable. One of the noticeable feature of his method is that arbitrary flow solver can be adopted for analysis part that gives actual pressure distribution.

He also programmed the design code named WINDES based on his theory.

Although his method can be theoretically applied for arbitrary geometry, WINDES can only treat wing or wing body configuration. Nevertheless, WINDES can be applied to more complex geometry utilizing the flexibility to the analysis code.

215

In this study, flow around Wing-Engine configuration including jet plume and air-intake was solved by a multi-block Navier-Stokes solver and its pressure distribution was employed for the WINDES's input.

This report is based on the paper presented at 27th Airplane Symposium (1989 in Japanese) of which Dr.Takanashi was one of the authors.

## Procedure of Aerodynamic Design

The wing shape is improved gradually by iteration of analysis and design step. The design code developed by Takanashi[2]names WINDES works to reduce the difference of between actual pressure which is given by the solution of multi-block Navier-Stokes solver, and target pressure which is set to satisfy required performance and constraint stated below. The flowchart of this procedure is shown in Fig. 1.

The details of the Navier-Stokes solver used in this work is stated in Section 4. The wing out of the nacelle only was designed in this study.

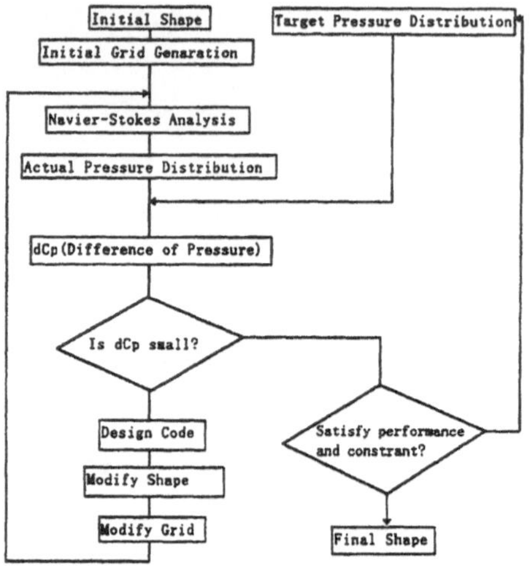

## Design Condition and Constraint

Fig. 1 Flowchart of design procedure

(1)Design Condition
Mach Number : $M_{DES}=0.75$
Angle of Attack : $\alpha=1°$
Reynolds Number based on MAC : $Re=2.02 \times 10^7$
Engine condition : Operating engine condition is set at inlet and outlet of engine nacelle as boundary conditions.

(2)Constraint
Wing area : $S=40m^2$
Aspect ratio:$AR=12$
Taper ratio: $\lambda=0.4$
Sweep back angle: $\Lambda_{25\%c}=3.107°,0°$ at aft beam (63%c)
Thickness ratio: $t/c_{root} \geq 0.16, t/c_{tip}=0.12$

(3)Design target
Lift Coefficient($C_L$) at design point:$C_{L\,DES}=0.32$

Natural laminar flow: Favorable pressure gradient before shock wave.

Off-design condition:

Target $M_D$. (Drag Divergence Mach number) conditions for off-design conditions are as follows.

For Speed overshoot

$M_D \geq M_{DES}+0.02$ at $C_{LDES}=0.32$

For Pull-up maneuver

$M_{DD\,PULL\text{-}UP} \geq 0.75$ at $C_{L\,PULL\text{-}UP}=1.3 \times C_L$ $_{DES}=0.42$

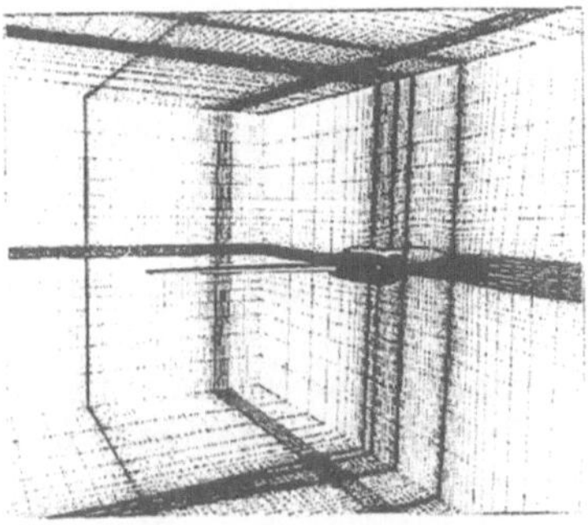

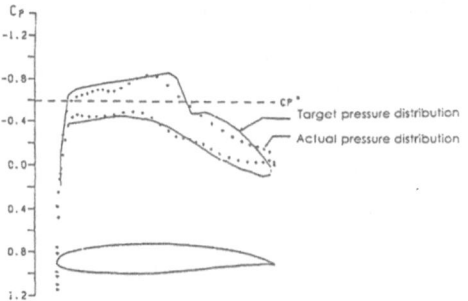

Fig. 2 Wing shape and grid on outer boundary and symmetric plane.

## Navier-Stokes solver

The multi-block Navier-Stokes solver developed by Sawada and Takanashi[1] was used in this work. It utilizes following physical model and numerical schemes,

(1)Reynolds averaged Navier-Stokes equation with Coakly's q-w two equations turbulence model,

(2)cell centered finite volume method,

Fig. 3 Target and actual (obtained by NS solver) pressure distribution.

(3)Chakravarthy-Osher pre-processing TVD scheme with Roe's approximate Riemann solver,

(4)symmetric planar Gauss-Seidel relaxation for implicit time integration.

The numerical grid for this computation is shown in Fig.2.

## Target Pressure Distribution

The target pressure distribution shown in Fig. 3 was adopted in this study. This distribution was set at all position out of nacelle aiming iso-bar design. The sectional shapes at a wing tip and close to a nacelle were fixed to the initial shape in order to keep wing thickness, because it was expected

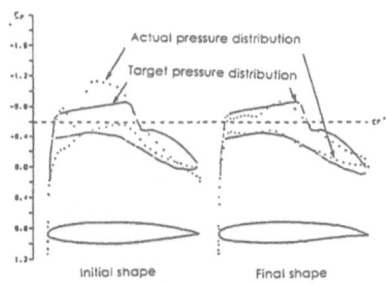

Fig. 4 Sectional Cp distribution at mid span.

217

that the resultant wing would be too thin if the same pressure distribution which had weaker shock wave was used directly on all wing position.

## Result and Discussion

The laminar flow region of the wing out of nacelle which was designed in this study was expanded because of expansion of favorable pressure gradient area before shock wave. On the other hand, the target $M_{DD}$ stated above was not achieved perfectly, however, it was better than the initial shape for both Speed over shoot and Pull-up Maneuver by weakening shock wave and moving shock wave backward.

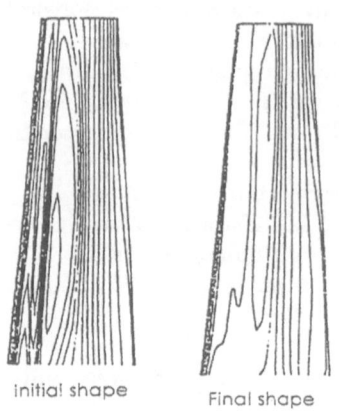

initial shape          Final shape

Fig. 5 Pressure contour on the wing

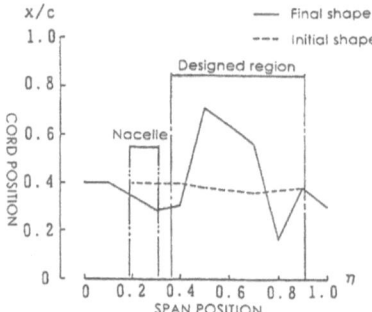

Fig. 6 Transition point distribution

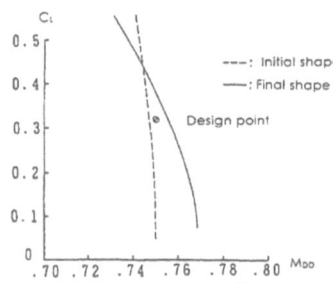

Fig. 7 Lift to $M_{DD}$ characteristic

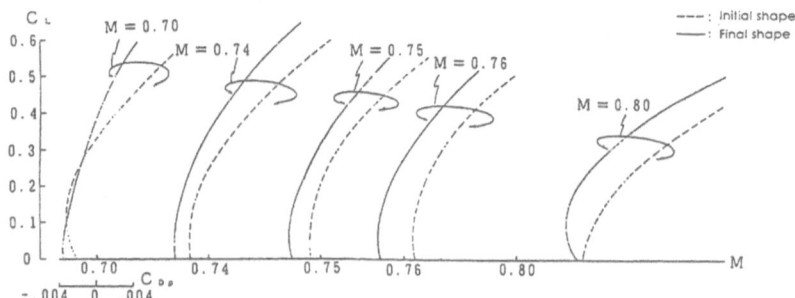

Fig. 8 L/D at each Mach number

See Fig. 7.

L/D performance was improved significantly.(Fig. 8) This was due to eliminating shock wave near leading edge found in the initial shape, weakening the shock at 60% code and achieving iso-bar design which realized the pressure distribution close to that of ideal two-dimensional airfoil at all position.(Figs. 4,5)

Moreover OEI(One Engine Inoperative) condition of twin jet USB configuration was simulated

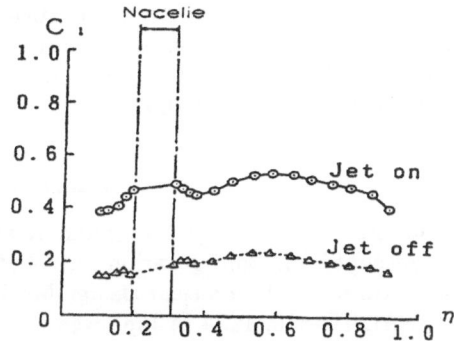

Fig. 9 $C_l$ distribution of Jet on/off condition

by giving appropriate boundary condition on inlet and exhaust of an inoperative engine in order to investigate the effect of jet plume on the wing which is unique characteristics of USB. It was found that the jet plume made significant effect on the whole wing. It is clearly seen comparing the result of Jet on and off computations.(See Figs. 9,10)

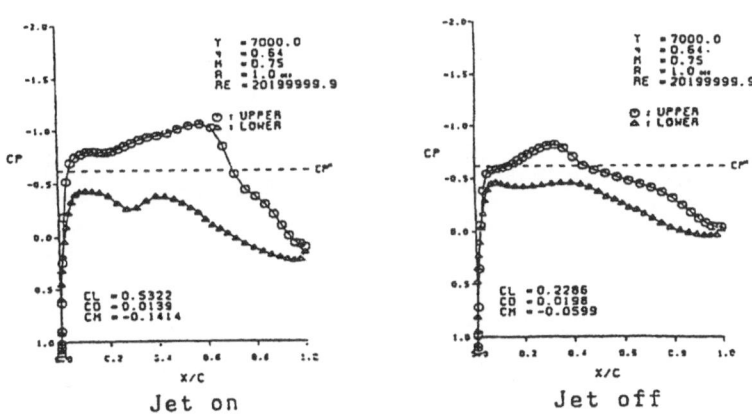

Fig. 10 Sectional Cp of Let on/off condition

## Conclusion

(1)More laminar flow region was realized and L/D performance was improved by the aerodynamic design using the inverse problem (design) code. Although initial target $M_{DD}$ value was not achieved, $M_{DD}$ was also improved. The favorable pressure gradient before shock wave which is appropriate for laminar flow tends to make shock wave stronger and to cause high wave drag, thus more study is necessary for high $M_{DD}$ and natural laminar wing design.

(2)The effective design procedure using the design code with Navier-Stokes analysis was established by this work.

## References

[1]Sawada,K.,Takanashi,S, "A Numerical Investigation on Wing/Nacelle Interfarences of USB Configuration", AIAA Paper 87-0455, 1987.

[2]Takanashi,S., "A Theoretical Design Method for Three-Dimensional Transonic Wings", NAL TR-830, 1984 in Japanese.

[3]Kamiya,N., J. Japan Soc. Aeronautics and Spacesciences, Vol.27, No.311, 1979, pp.627-637 in Japanese.

[4]Kamiya,N.,"On the Drag Divergence of Two-Dimensional Airfoils at Transonic Speeds", NAL TR-299, 1973 in Japanese.

[5]Eggleston,B,Poole,R.J.,Jones,D.J.,Khalid,M., "Thick Supercritical Airfoils with Low Drag and Natural Laminar Flow", J. Aircraft, Vol.24, June 1987, pp.405-410.

**Brief Instruction for Authors**

Manuscripts should have well over 100 pages. As they will be reproduced photomechanically they should be produced with utmost care according to the guidelines, which will be supplied on request. Figures and diagrams should be lettered accordingly so as to produce letters not smaller than 2 mm in print. The same is valid for handwritten formulae. Manuscripts (in English) or proposals should be sent to the general editor, Prof. Dr. E. H. Hirschel, Herzog-Heinrich-Weg 6, D-85604 Zorneding.